T0127704

Ocean Acidity Climate Shock

Science Fallen and Risen and the Art of Magic Investment

Chondrally

authorHOUSE®

AuthorHouse™
1663 Liberty Drive
Bloomington, IN 47403
www.authorhouse.com
Phone: 1 (800) 839-8640

© 2018 Chondrally. All rights reserved.

No part of this book may be reproduced, stored in a retrieval system, or transmitted by any means without the written permission of the author.

Published by AuthorHouse 03/20/2018

ISBN: 978-1-5462-2950-6 (sc)
ISBN: 978-1-5462-2949-0 (e)

Print information available on the last page.

Any people depicted in stock imagery provided by Getty Images are models, and such images are being used for illustrative purposes only.
Certain stock imagery © Getty Images.

This book is printed on acid-free paper.

Because of the dynamic nature of the Internet, any web addresses or links contained in this book may have changed since publication and may no longer be valid. The views expressed in this work are solely those of the author and do not necessarily reflect the views of the publisher, and the publisher hereby disclaims any responsibility for them.

The Forgiven:

Harjit Sajjan, Justin Trudeau,Elizabeth May,Colin Wilson,Ed Jernigan, Ged McLean, Tim Topper,Rajesh Jha, George Soulis, Barry Wills, Prof Brundrett, John Robinson, Gordon Savage, Gord Doig, Peter, Raj Pathria, Anu Pathria,Prof. Vanderkoy,Ken Leslie,Vanessa Doig, Bruno Forte, Glenn Heppler, Prof. Koncay Hussein, David Harding, Ashok Kumar, Scott Allen, Paul Wesson, Tom Lee, David Fleming, David Kantor,Ian Stewart,Dr. Girgla,Dr. Habib, Dr V. Van,Richard Feynman,Canny,Kate and Austin,John Constant,Maurice Constant, Leon Cohen, Prof. Mallat, Ed Witten, John Wheeler, Walter Duly, Mike Fich,Michael Bronskill, Mark Henkelman,Graham Wright, Claude Nahmias, Andy Hulme, Hank Chan,Tom Lang, Roger Webster,Ron Trisnan, John Watts, Dave Ibbetson,Jack Currie,Brenda Currie,Bob McDonald,Andrew Wong,Louis Wong, Chris Pratley,David Atherley, Bruce Patterson,Bill Marchant,Ray Manninen,Roel Mahatoo, Ellen O'Dwyer,Bob Roach and family,Rob Johnson,Simon Rawlinson,Roland Tanglao, Theresa Salter, Yalan Kwon,Peter Gill, Iain St. Martin, Susan Noppe,Romany Woodbeck,Sandra Di Diomette and Jill, Patrick, Sarah, Ewan, Jenna, Scott and Jamie, Ken and Jen Fleming, Slawo Wesolkowski,Anna Zisniewska, Lindi Wahl, Naomi Tague, Bob Bowerman and Chris Robson, Niall Fraswer, Kish Hahn, Mark and Patti and Riley Murphy, David Baldacci,Bruce Rieger, Helmut,Kiyo and Tamiko and Alex and Emily Tabuchi,Ken Adkin, Chris Barlow,Keith Patterson,Iain McDonald,Pierre Lemieux, Jacqes Hegel, Moussa Baalbaki, Youssef Sabeh,Jae S. Lim, Jan Narveson, Keith Hipel, Bill Lynch,Michael Palin, Alexa Salvatori,John Cleese, Leonard Susskind, Stephen Hawking, Deepak Chopra, Oprah Winfrey, Cervantes, Aldous Huxley, HG Wells,Orson Scott Card,Harper Lee, Arthur Miller,

Joseph Heller,Orson Welles.Ron Greidanus,David Harding,Christine Harding-Lane-Smith, Noam Chomsky,L Ron Hubbard,Brittany Malm,Robert Chin, Sarah Jane Allen,Arthur C. Clarke, Isaac Asimov, Chris Jubien,Steve Mann,Jonathan Allore,Jonathan Blake,Harry and William Windsor, Elizabeth Windsor (nee Gotha), Charles Windsor, Margaret Windsor, Kate Windsor,Megan Markle,Taleb,Sornette,Benoit Mandelbot,Andrew Heunis, William Jennings,Al Otten and Gabriel,Iain Otten, Eric Otten, Luke Otten,John Murray and wife,Stephanie, Ian Veenstra and Roly, Brian Yealand,Charlotte Beer, Charles Beer, Mary Sanderson,Evan Siddall,Steve Tancoo and sister,Dave Kirby,Lachlan(Larry) W. Smith,Joseph LEsperance,Jane LEsperance, Lygia Daudet,Doug Hastings,Duane Hatcher,Avicenna,Archimedes, Newton, Leibniz,Richard Feynman,Tarmi Claussen,Donald Knuth,Alex Ashenhurst,Jim and Kathleen Christie,Grace Schmidt, Peter and Mary Roe,Mark Ezard,Richard Z'Graggen and Edna, Janet Mason, Robert Z'graggen,Mary Beth Kyer, Claudette Johnson, Joe Bluestein, Chris Nippard, Joey Brake,David Suzuki,Nick Lane,Stephen Lane-Smith,Andrew Lane-Smith,Gord Jans,Stephen Scott, Thomas White,Carolyn Beeton,Matt and Sybil Larmour and David Larmour,Elizabeth Brauer, Cynthia Pauli-Ball,Grace Colon,Robert Allen,Greg Dee,Alan Greg,Kevin and Gary Gillespie,Malcolm MacLeod,Sandra Lovegrove, Lino and Sylvana DesGasperis, Sheynal Saujani,Beth Mcandless, Mary Siddal, Tom and Mary Pettingill,Gwynne Dyer,Len and Ann Smith, John and Margaret Anderson, David Anderson,Iain Anderson, Mr Leggatt and Ishbel and Christine and Ewan and Calum,Mark and Tara Bostock,Ernie Lewis,Doug Wallace,Richard Zeebe,Dieter Wolf-Gladrow,Jeff Wood, Dianne Lough,Louis Dajenais,Rick Dajenais,Randall Bier,Mr. Cotton, Mr. Ogden, Mr. Bassett,Donald Toon,Peter Garrioch, Ms. Larsen, Art Jattan, The Winskis, Ayako,Kiriko, and Azuko and Suzuko,Phil Bryden,Sharon Walt,Dianna Del-Bel-Belluz,Mitali De, Stephen Birkett,Tracy Feltham, Catherine Booth,Ron Nielsen, Benj Fayle,Iain McNicol, Tess Gates,Tom and Laura Gates,Clive Allen, Derik Hawley, Karen Lucas, Alain Lucas,JP Pawliw,Kathy and Dave Bowman, Laurent Ramillon,Jean-Luc Sune,Jacques, Johnny Murray, Monique Monnier,Dianne Beausejour,Ingrid Bongartz, Jean-Luc

Getti, Christophe Lienhard,Steve Roach,Duane Wilson,Tony Iarocci, Tony Edmonds, John Owen, Margaret Owen, Glynn Owen, Josh and Rebecca Owen,Tony,Mrs Kewley,Mr. Sturm,Jerome,Stephen Joyce, Kevin Moon,Laura Kingsbury, Laurette McDonald,Pauline Newman,The Meddaughs,Tammy and Tim and Andy and Jason,Rick and Ed Meddaugh,Ashok and Valerie Kumar,Mr. Moore,Bob Hamilton,Amanda,Michael Cummins,Amanda Pacheco,Darshan, John Wolfe,Tom Cochrane,Stan Fogal,Poincare,Pasteur,Schrodinger,Andrew Wiles, David Ash,Lawrence Pilch,Phil Willow,Jim Pianosi,Jamie Champ,Andy Dobson,Jim Parent,Mike Hughes,Marty and Mary Hughes,Rosemary Vickers, Rosemary Obadia, Adrian Obadia,Derek Lane-Smith,Mrs Pilon,Mr. Monohan,Dieter Turowski,Bruce Hulme,Richard Jordan,Tino,Corey,Paul Dilda,Cassandra,Harmeet K. Bami, Ruth Groome-Kurtz,Nick Hatton,Ron Hatton,Rose Van-Klanski,Rodin,Brancusi,Henry Moore,Wigi,Ansel Adams,Annie Liebovitz,Rembrandt,Matisse,Monet,Manet,Hugh Moorehead,John Moorehead,Peter O'Toole,Tom Hanks and Sean Connery,Daniel Radcliffe, Emma Thompson, Rupert Grint,Izzy Simpson,JK Rowling,Martin Buber,John Williams,Segovia,Eric Clapton,Celine Dion,Carolyn Brodeur, Josee Byron,Andre Chevrier,Peter Majerski,Peter and Peter Kuryllowich, Betsy and Kara Kuryllowich, Paul Wesson,Loreena McKennit, Ed Sheeran,Wendell Berry,Langdon Winner,Russell Peters, Will Smith and Rajesh Jha and Sigourney Weaver and Keanu Reeves and Rob deKemp,Rick Baron, Steve Stein,Jim Quick and Dianna Rigg (Emma Peel) and Tom Tsai and Geoffrey Hinton and Francis Ford Copula and Leonardo Da Vinci and Leonardo Di Caprio and Google and Microsoft and Kathleen Wynne,Brian Mulroney, Brian Milliere,Polkinghorne, Peter Barrow,PC Trulson, Pope Francis and the Archbishop of Canterbury and Dave Small and Asad and Shaymaila Ali and Frank Capra and Henri Cartier-Bresson, the Matrix and Avatar, Spock and James T. Kirk,Data and Picard,Arnold Schwarzenegger,Dalai Lama,Aubrey de grey and David Copperfield and Penn and Teller and all the Good Witches and Wizards and all the Trappist Monks and Life, The Universe and Everything. There has been a lot of Justice! It's all about Polynomials. Poly – Many, Nom- Names

Dedicated to John Raulston Saul and Nelson Mandela and Eleanor Roosevelt and my Family:

Contents

Introduction

Is it too much to ask to feel some Joy in life and some Love and Pleasure and perceive some Beauty and Melody. What do we have to worry about? The Truth is the ultimate victim, too much truth is deadly and no one group possesses the whole story, even the whole story of a person's life might never be known in its entirety.

Cervantes knew that! Don Quixote was written to escape the Spanish Inquisition, A story of a mad man that no one could pin down or blame and that everyone laughed at and kind of loved or at least liked.

He was entertaining. Much like Chaplin in his own time!

The Dalai Lama interviewed recently by John Oliver in Dharamsala, India expressed that hardliners on the left and right were missing a part of their Brains. He said, usually the Brain has the ability to create common sense. And the Hardliners are unable to do this. They also have trouble with perspective, laughing and smiling, and they grimace a lot. They are worried about what people think of them and want to control people so that they obey. It is antithetical to freedom! Nelson Mandela and Chaucer wouldn't have liked the hardliners and that is to Chaucer's credit. Chaucer reacts instinctively and negatively towards hardliners. But the question is, can he create Common-Sense?

https://www.youtube.com/watch?v=bLY45o6rHm0
https://www.youtube.com/watch?v=xecEV4dSAXE

https://www.youtube.com/watch?v=q8U7GlcQFDM

https://www.youtube.com/watch?v=jFWcSVPzxWI

On the contrary, Chaucer makes so much sense that it is mostly Nonsense! His Truth doesn't exist for long and what comes out of his mouth or his Twitter account next can be hilarious if people take it too seriously! Hooray for Alec Baldwin, He probably saved Chaucer's life!

It is the Post-Truth Era! Everything seems like an illusion! George Orwell warned us about it! It was Orwell's book 1984 about a Totalitarian hell on Earth where pain,control and fear ruled that might emerge, where all resources are used up by the state, and the people are all under mind control with no free will or access to real news. Huxley also warned in Brave New World about genetic engineering, the alphas, the betas, and the deltas and epsilons….and how everyone would be bred to be slaves to the state and addicted to 'Soma' and pleasure as well. Now we know it is a hybrid between the two.

All the illusions are really all because of the blind spot in the eye and because of ambiguity in language (all the puns and double and triple meanings in English and other languages), and also because of the rods and cones in the eye. At night time, when we are using our rods, colour eludes us, and sharp definition eludes us too. Often motion and shape are miscategorised by our brain, and we may see angels or demons in the dark. The blind spot is the ultimate clue, it is the whole basis of magic and misdirection. If you learn the trick, no one can ever see how you did it. It is also what all criminals do. There is a fine line between a good magician and a criminal. It depends on their intent. If it is meant for entertainment or something benevolent like a white lie or to steal something that is deadly to someone else, so they can be relieved of the torture, then it is good, and they are not criminals; if they intend to steal something good from someone so that they can profit or demean the other, then it is criminal. Everybody blinks; how much time passes for other people when it happens no one can always be sure. Only a scientist can make it ok. It really is quantum mechanics, time and frequency, the Uncertainty principle, and General Relativity, each person has a light cone of perception and scientists are connected to satellites so they know when a space-time anomaly occurs.

Extra Sensory Perception and Telepathy are real because of radio waves, microwaves and telecommunications. Channels and frequencies and codes and cyphers are real too. Scientology can help examine the moments of unconsciousness (engrams) when some time or energy was stolen, but they charge too much themselves and turn their believers into missionaries addicted to their goals, it is not free will either.

The perspective that everybody was a kid once and knows that there are playground bullies, and meek and timid and shy people that are too afraid to talk or stand up for anybody or stand up for themselves should tell us a lot!

If the meek and timid and shy all vote in secret, the bullies will be surprised!

Perhaps they really will inherit the Earth! But that might be uproarious, because it looks like the Oceans might die around 2040 to 2060! Or at least they will be under serious threat from Ocean Acidity Climate Shock. The Calcium Carbonate buffer will suddenly reach a new equilibrium around 493 parts per million (ppm) of CO2 in the Ocean surface water by 2052. And the pH will suddenly change by up to .55 pH units, first more Acidic, and then more Basic and then more Acidic again in an undulating oscillation that lasts a few years. The phytoplankton and algae might not be able to survive the pH and temperature changes, unless we seed the Ocean with genetically modified phytoplankton and algae or selectively hybridized phytoplankton and algae and krill that have been shown to survive pH and temperature changes of those magnitudes and pollution. Scientists are working on this problem right now and NASA has special satellites monitoring the phytoplankton levels in the Arctic, the most delicate and intimately connected ecosystem that indicates how the fish species are surviving in those waters. If the phytoplankton and algae and krill die, it means half the world`s oxygen supply and is the base of the food chain in the ocean. Disaster!

Oceans losing their breath! Warmer waters hold less dissolved oxygen, and algae blooms due to fertilizer runoff are starving the ocean of oxygen and causing oxygen dead zones when they decay!

https://serc.si.edu/media/press-release/ocean-losing-its-breath-heres-global-scope
https://www.theguardian.com/environment/2018/jan/04/oceans-suffocating-dead-zones-oxygen-starved

Climate Change drives Collapse in Marine Food Webs! As phytoplankton disappear due to pH and temperature changes and it affects the base of the food chain in the ocean. 40% of phytoplankton has disappeared since the 1950's. Fish stocks have been declining ever since due to lack of food supply and overfishing:.

https://phys.org/news/2018-01-climate-collapse-marine-food-webs.html

People have been studying this problem and felt somehow they had to tell the world about it and warn them. But under Chaucer who felt it was all a Chinese Hoax, and there are some who only could read balance sheets, only Merkel and Trudeau and the British and other Europeans and even the Chinese seemed to care but not the hardliners. But there never was a TV show about the Calcium carbonate buffer, or about the Calcium carbonate buffer becoming saturated later on about 2232 at 1329 ppm CO2, which was a much more serious problem for the Oceans than the phase I Calcium carbonate fluctuation if that line was crossed, for sure the Oceans and life in them would not survive! Even CBC the Nature of Things didn't do a TV show about it! When the Italian Seismologists predicted an Earthquake in a Mountain town, they went to the town and warned people and were kicked out of town. After the earthquake happened there and people died, they were put in jail, and they were blamed for it. No wonder no one wants to do a show about it.

Were we all going to collectively commit suicide by ignoring the problem? Effectively guaranteeing our grandchildren enormous debt and no planet to live on. Essentially Hell on Earth! Were we all just going to be Nero, fiddling and partying while Rome burned!

Chaucer just isn't either smart enough to understand it, or is too much of a kid to face it like a grown up or is actually going to change his mind at the last minute when he has no other choice. The US Women are all furious with the misogyny of the Administration and it looks like a real highway super crash that everybody and their kids is rubbernecking at. He has turned us all into gawking spectators at a spectacle and everybody is horrified. The women forget that many of them are guilty of blackmail and misandry! It doesn't sound like like or love! Many of the women sound like hardliners and they sound hostile! It is really annoying that some of the Lesbians were man-haters (misandry - not all by any stretch of the imagination) and Gays tried to turn people into homosexuals when they really rather would not for political clout and gain. Some people had no choice but to become Gay or Lesbian just to survive the military dictatorship. Some of them have very high intelligence, from a life time of insight, education and persecution. Personally not Gay but can appreciate their experience. The ones who hate for a political or social cause: Did they ever really love anybody or their children? For the Men, was it really all about power and money and land and Oil. Their misogyny and attempts to grab power without justification or merit are really petty and annoying and downright wicked. Also the desire of some women to have absolute power and authority over men is unrealistic, as it is also unrealistic for men to want absolute power and authority over women as well. We need Diplomacy and we need to make agreements. Sometimes the Deals do become too toxic (talk-sic) and poisonous for anyone and need to be stopped absolutely. We do need to listen to and talk with the young people.

When dealing with a mass audience, Generalizations are necessary to stay sane and safe. However stereotypes are only that, no one person actually fits all the stereotypes and eventually everyone falls through the cracks of the generalizations; people usually learn, one way or the other, the stereotypes don't actually last all that long; that's why politics always has to be remade with new generalizations or qualifications and many exceptions. On a personal level, Generalizations are fundamentally stupid, when you hate ALL men (misandry), or hate ALL women (misogyny), this leads to ridiculous interactions with friends. It may be a phase many jilted or hurt middle aged people go through!

https://en.wikipedia.org/wiki/Misandry
https://en.wikipedia.org/wiki/Mysogyny

Every single person is different and unique, and every single one has some virtues and some flaws. It would be nicer if people started to notice the virtues instead of focussing on the flaws all the time. People really do want to get better and focus on their strengths. I understand sometimes a good friend has to point out the flaws to them in a nice and diplomatic way so that they can improve themselves on their negative side. But don't forget to mention their virtues too and let them know you might be telling them because you love them!

Ultimately, we may have to all accept them just the way they are: Trying to change them all the time by education can be very tiresome especially as they get older and less able to adapt to changing circumstances. Lifelong learning is a goal we need, but we do need to develop character too or else there will never be any fun or entertainment. Noone can ever really be perfect about everything. People all really do have to go to the Toilet. It might be wiser to build a good,clean,safe and healthy Toilet that people need than focus on superficial wants like an expensive car. It is ultimately worth more to people. A Sanitation Engineer is worth more than a Car Salesman. Noone can know everything about everything. Authority eventually becomes obsolete, it's punishing and not rewarding. People really don't like it very much and that is why. They really appreciate a clean Toilet. You can live without an expensive car, but you can't live without a clean Toilet. Maybe a Car Salesman who is also a Sanitation Engineer might be worth more still. IT is all about Sulu and Chekov. I prefer Moliere. Parody and Satire are a bit too cruel and guilty amounting to blackmail and not so original, but essential sometimes in self defense, and it can really illustrate a point. Ingenousness wins more often but is harder and rarer to achieve.

Even if you know an absolute truth and everybody is wrong, do you really want to be Socrates and be assassinated by them because of it....No, we should rather focus on living and loving, but if our home Earth is not going to survive because of inaction, it's common sense to do something about it: Does Chaucer realize that? Its also better to err on the side of caution. Che Guevara made a huge mistake, he only lived to 39 and he was personally sacrificed for the cause. Some of his sentiments were essentially correct however he was too violent. Albert Einstein did the right thing, he left Nazi Germany for Switzerland then America and lived to a ripe old age peacefully, comfortably and with humour, accomplishing a lot more than Che Guevara ever could; not everyone can do this and a lot of people are inclined to stay and fight. It might be wiser to connect with firm but peaceful intelligent people who can smile about the innocence. Its always wiser to outsmart those who want direct conflict, or who want to directly punish just because something seems unfair, they might not really know the whole story. Exceptions do have to exist. You cannot treat a handicapped person the same way as a Sports star. Remember Sports is always trying to steal the ball from the other players. It is based in open and admired theft. In Sports the weak don't often have a chance. People do get injured in Sports. I'll stick with lawn bowling, Croquet, Golf, Cricket, Baseball, swimming and Pool and treat old people with respect. We need to avoid depravity and beware of the crucial difference between love, lust and attraction, repulsion and what some people try to do to others in their sleep.

Please review some of the following material to see some of the broken problems we do have to deal with in this Century:

21st Century - Broken Century

https://www.youtube.com/watch?v=pzqAnM0-CcE
Naomi Klein: This Changes Everything! https://www.youtube.com/watch?v=VthS0PDbiP8
https://www.youtube.com/watch?v=Rqw99rJYq8Q
https://www.youtube.com/watch?v=sKTmwu3ynOY

Solutions to the Climate Crisis:

One solution to the CO2 crisis is to use to use a hybrid Syngas-Solar engine that emits no CO2 and uses no net oxygen. The engine captures the water and CO2 from the waste gases and catalytically turns them back into the original fuel. This engine could be incorporated into a car, a truck or a power plant for electricity. This really is an option, see later chapters for details. The oil companies would hate this but it could save the planet and the oceans. All the technology exists to accomplish this within a couple of years. A second option is to use a water-lightning engine based on Sonoluminescence. It can generate lightning in water with sound waves due to resonance and the lightning current could be captured by a super capacitor battery that charges to high voltage in nano or milli-seconds. Then an electric engine can be run off the electricity at 80% efficiency. The sonoluminescense engine is a bit too far in the future, maybe 10 or 20 years away so we are best to stay with the hybrid engine mentioned first. Electric cars are problematic in the US and China because they get their electricity from coal plants and natural gas power plants that emit CO2. This is net almost as bad as gasoline and diesel engines for the environment. Clean coal CAN exist, but it seems likely they are not actually doing it because they don't believe in the Science and it would cost them more in the short term. In the long term it would cost them more when they see the Earth is in peril. One positive spinoff is that Electric cars can reduce smog in cities. The range of electric cars is improving with better batteries, but we are still waiting for a breakthrough in fast charging super capacitors maybe with graphene in them and generally their range is disappointing.

Clean Coal as a concept definitely exists, however it generally is not being practiced due to its high cost. If the true cost to their grandchildren of the environmental disaster were factored in it would obviously become economic for them to implement it. But they are too short sighted to see the true cost. The coal lobby is destructive and short sighted. They probably don't understand the science as well and they are probably too ignorant. The future will describe them as Darwin Award Winners. China is lucky, under a Communist government they really can implement 30 year plans, the west for the most part is bound by 4 and 5 year plans, occasionally a party gets a mandate for about 8-10 years. Nothing in the west is certain beyond that, and that's why it is too short sighted. The long range plans never get funded properly in the West.

People are really afraid of Science, and no wonder, cancer rates have been increasing due to chemical and physics pollution and active targeting.

People are afraid of the microwaves and not a few people want to wear aluminum foil hats but are afraid they will get Alzheimers from the Aluminum. The synchrotrons and CERN really scare people. It sounds like the physicists only care about tiny particles and astronomical budgets, and people are afraid they are shooting the tiny particles at them if they don't go along with it and they feel blackmailed by them. They feel it and they don't have the vocabulary to even talk about it. The politicians won't ever say anything about the Oceans or the particles and we know the TV networks can save people from the particles if you are lucky enough to be one of the Elite. People are afraid that a part of their Brains will be affected if they get involved. They can feel the powerful rays, and they don't' want to get caught in the middle of a media war. They feel that a lot of things just aren't right for them or their families and they feel like they have no effective voice in parliament that can get things done to undo the injustice that has been dealt to them.

Only Al Gore and the Green Party candidates will ever talk about the pH of the Ocean. And only briefly. To be fair the Liberals and Democrats do fully seem to understand the dire situation but feel there are more important problems. President Obama knows. Premier Kathleen Wynne has heard about the Oceans! The discipline of some good Conservatives is really welcome, thankyou Mike Chong(Conservative). Thankyou Justin Trudeau (Liberal) and Harjit Sajjan (:Liberal) and Jagmeet Singh. (New Democratic Party - NDP) and Elizabeth May (Green Party).

The normal citizen is just fearful and is too poor to worry about Climate change and just cares where they are going to get the money to pay their next Hydro bill and pay the rent and put food on the table while still being able to watch a bit of entertaining TV like X factor or American Idol or Big Brother or Survivor or Game of Thrones or the Originals or Britain's Got Talent and still be able to sleep at night and they want pot and liquor, I think they need to learn some music, some opera, and something a bit more refined than sports, like fine dining or Cricket, or a game of Euchre or Crazy Eights (it's always amusing)! Keep in mind that many people got into drugs in the first place because they were really afraid of Nuclear effects on them, and for some, it was the only way to avoid it. A little compassion goes a long way. A lot of people chose to be Gay or Lesbian because of it too and some people were forced that way for their own good. Being Gay or Lesbian might be a better choice than being prematurely dead due to drugs or the military or Religion. Most were born that way and it wasn't a choice for them, it was simply against their nature to be any other way. Their brains and bodies were differently connected from a young age. They have a joie de vivre too and a sense of humour, and are more often peaceful people. The persecution isn't really justified. They know what the science and military and drugs tried to do to them. It is 2018 now, and people do have to accept that they exist and live. It really is harder than young people think to get married and raise kids. Religion can help if you are heterosexual especially. When people are young many experiment with homosexual or bisexual, and it is not abnormal. We do have to be careful about safe sex with condoms, and certain other acts. Exposure to bodily fluids is risky and many people are not careful enough. If you want to raise kids or be a Teacher or someone involved with the Public, Casual Sex is not recommended. Many people have big hang ups about it, as we have recently all seen in the news. Even asking a woman if she wants to 'sleep' with you is almost always assumed to imply intercourse. People really are too Freudian for their own good. I'm sure Steve Paikin is innocent, he probably just wanted to hear her at night. Carl Jung was more on the right track. Everybody needs it, it makes everybody stupid, people gossip about it, and rumour ruins peoples reputations more often than not. It is supposed to be private between two people, there may be too many threesomes and foursomes. If it is something good, word spreads. If it is something bad, word spreads. Neutrality might be the only sane response. It is Peoples' hatred that really gets them in trouble when they really don't get what they want; it really still is very childish. The kids aren't learning at home from their parents, society won't let parents teach much; its only what they learn on the computer and TV and from their friends and school. 'God only knows' what they have really been exposed to. Nobody could get it correct, and nobody said it right, and everybody was too frightened to really talk about it. It was a lot of Freudian suspicion. The understanding of Carl Jung's book 'Man and his Symbols' and Semiotics is really important. We also need a book about 'Women and her Symbols'. Climate Change might just be too heavy an issue for anybody to deal with on a personal level with all these other problems we are all experiencing. We do want to enjoy ourselves and prosper. It might really be the TV and Mass Media companies and Medicine and the Military that are abusing them, both men and women.

Some Religious people really care, but most of them still don't know any of the Science and even some Scientists will deny that unsafe nuclear is happening, saying only that the levels of radiation are safe never mind targeted at certain individuals as a weapon by the State. Does anybody know any of the Engineers that work at the Nuclear Power Plant? I've only met one or two in my whole life! And it seems they only have a half-life outside of the public spotlight! I believe there is a talking ban on everybody that works at the Nuclear Plants, nobody ever hears anything about it, except for very carefully edited columns in some of the newspapers, with practically no public discussion!

Observable and Controllable are not the same thing. And if you have a system in control, in one`s own Brain, it might not be wise to observe it. Because if it is fine, and small structure, observing it can change it, by Quantum Mechanics and the Uncertainty Principle. You can even break fine processes in your own Brain that can take years to regain or rebuild if they can be rebuilt at all, and the chaos that ensues might result in Mental Illness for a while or indefinitely until such time as in the future when they can repair the process, even reading and writing can be affected! And it is a very fine study of phonetics, and sound as phonons (sound particles) and phonemes (fricatives, plosives and diphthongs)

https://en.wikipedia.org/wiki/Phoneme

that can save you from true photons that are dangerous. Sound is ultimately linked to meaning and a sound mind and sanity. Brilliant might be overrated and dangerous. Light can be dangerous. Learn to listen more than speak. Most of Society is Observable but not Controllable. That's why politics can be such a phallacy and it makes a lot of people Cynical. Most people forget what happened the next day and they even forget who people are. There really has been too much abuse and war and manipulation for profit. Can they really love them? Do manic people, Presidents and workaholics understand that people need to rest? There really is too much mental illness, it is funny and sad at the same time! They should read Stephen Pinker's 'How the Mind works' and 'The better Angels of our Nature'.

If God or Yahweh or the 1 trillion names for God or Aliens exist and they can change the laws of Physics and Chemistry, equivalent to Magic, it might be our only hope of survival. If God actually exists as many people believe, then it can control Aliens as well as God made them too. Let's hope the Pope and the Archbishop of Canterbury and a Well Meaning and Good Intentioned and Benevolent and peaceful Rabbi and Imam and imamah

https://en.wikipedia.org/wiki/Rabbi
https://en.wikipedia.org/wiki/Imam

and the Dalai Lama can make Susskind and Hawking face up to the Anthropic Principle and the Goldilocks Zone. I really think it has something to do with Anthropology! An apology for Man.

https://en.wikipedia.org/wiki/Anthropic_principle

If there is no Faith it looks just like randomness or chance. We might have to admit that randomness or chance still exist but Einstein and the Religious believe it may be an illusion of God especially if there is no Faith and admit that the 2nd Law of Thermodynamics is False for Open Systems. Negative Entropy does exist; Governments, Institutions and Companies rely on it. Polkinghorne might well be correct, the Quantum World and the Resurrection might not be mutually exclusive. God and Science can be reconcilable. Science isn't inherently evil and neither is God, Up until now Science sounded more intelligent, but that is also an illusion of ego and vocabulary, even simple people know how it feels. Science and God can save or destroy, much like any human being. The real question is what are you trying to save or destroy and why? How far into the future are you planning, and how far into the past does it agree with? If you are building a bridge to a stable and benevolent future, what is the problem?

The Architect and Civil Engineer need to listen and speak! The Architect of the Human Brain needs to be involved in the discussion!

To Sum up, The Oceans are on the verge of dying. If they go, there will be a chain reaction onto land first with the Oxygen, then with the food chain. It could spell an apocalypse for the whole planet Earth for all living species. It doesn't sound good. Only Science or Art or Divine Intervention or Alien Intervention or Magic could possibly save us. I am placing my bets with all five, as they are the only understandable and repeatable disciplines that could actually make the difference for sure. I really do believe in the story of Harry Potter. All the kids will remember. And the Story of George R. McDonald ('The Goblin and the Princess'). Lord of the Rings by Tolkien. Artists can really create Beauty or Ugliness. Some Aliens might exist among us from time immemorial. Christ himself might have been half-Alien. Religion, Aliens and Magic are not always repeatable or known. They are not always a sure bet. Science, Art and some Magic and some Religion are repeatable demonstrably. Religion is what created the Schools and the Hospitals, along with Charles Dickens. Some Magicians can always repeat the trick. They will never tell you how they did it. Once the secret is known, all the mystery goes away and all the wonder and they are all disillusioned… because it was only a trick. And their description of what the magic was may be a deception itself to make the curious leave. Only Science and Art and Magic are demonstrably repeatable and educable to those with the Talent. Scientists and Artists are essentially Magicians and Wizards and Witches. It's not really any different than Magic. Nobody else can do it either. They are looking for people with Talent.

Gwynne Dyer, an expert in warfare is turning his mind to scuba diving and saving the coral reefs with assisted evolution and geo-engineering!

Dyer: Coral reefs, assisted evolution and geo-engineering

https://www.kamloopsthisweek.com/dyer-coral-reefs-assisted-evolution-geo...

GeoEngineering could hold back climate warming, but what if the dam bursts?

http://www.cbc.ca/radio/quirks/breathing-spreads-the-flu-what-happens-if-we-stop-geoengineering-your-eyes-are-pointing-your-eardrums-1.4502554/geoengineering-could-hold-back-climate-warming-but-what-if-the-dam-bursts-1.4505302

Ecologist rates Thai coral reef decay rate as alarming

https://www.bangkokpost.com/news/general/1403638/ecologist-rates-thai-coral-reef-decay-rate-as-alarming

There is a book called 'Religion and the Decline of Magic' by Keith Thomas. It might be good to remember the lessons of history and science. Magic brought con artists and fake healers, and snake oil salesmen...... don't forget, at least science tests its treatments statistically and based on sound biochemical and biophysical principles. Where it went wrong was that some of the diseases did not exist before the Nuclear Age and they caused real suffering. We need to remember the lesson of the film 'The Third Man' with Orson Welles about a con artist businessman who sells fake penicillin to orphanages in post-world war II Vienna. We also need to remember the radio broadcast of the War of the Worlds by HG Wells, starring Orson. If you want to survive the best bet now is to focus on sounds and phonons and phonemes and a sound mind and vocabulary, and learn to listen first then speak in self defense. And learn to feel the pains and pleasures of your body. If it feels too good it's probably too good to be true (like heroin or cocaine). If it feels too bad, it's probably not worthwhile, but you had better learn a lesson from it and REMEMBER. It all sounds like an Alien conspiracy to steal all the resources of Earth. I hope God and Justice exist. Immodium does work, so does insulin, so does ex-lax, so does soap and water, so does brushing your teeth with fluoride. What is wrong is that there are criminal gangs out there who are working to poison all the medicine. All the side effects are not good. There are problems in the pharmaceutical industry because of it's for profit nature, some of them may have lost the Hippocratic Oath, and there are problems catching the criminals that are poisoning the true medicine. The focus on prescribing pills is a bit sick in itself. They have to trust the Naturopath as well. The GP feels their pain and takes a proper history of the person.

Some Physicians do have the Hippocratic Oath still, and appendicitis is real, so are kidney stones, so is a broken leg. Physicians can help fix these things properly. The whole thing was caused by a bug which is not the true assessment, there are many social causes for disease! Read Plague Time by Paul Ewald, and The Limits to Medicine by Ivan Illich. There are Drug Lords and crime syndicates. The truth isn't so simple. There are some very good social scientists and physicians. Remember Generalizations are not good, and we are all working in an imperfect system with limited time and resources, if you notice something is not correct you need to notify the physician or naturopath or the astronomer!

Astronomy and AstroPhysics knows what's happening ahead of time; All the satellites worth knowing report to them!

The illusion of Aliens is probably due to collusion between the mass media and the military and computers and Religion. During the cold war, it was imperative to make the enemy believe we had access to alien technology and thus put fear into the enemy and demoralize them. The mass media and the military can control the visual and auditory cortexes of almost anyone, and make them hear and see whatever they wanted them to see and hear. It is not God but advanced NeuroTechnology.

Neils Bohr based his insight into physical chemistry and quantum physics based on a metaphor from Kepler, but focused it inward instead of outward. Outward is safer. The focus direction is what created the big bang of the nuclear explosion ultimately. Everything was timed to exact detail. The Manhattan project used all their mathematical and physical prowess to create the explosion. It's ultimate intent was to destroy life and liberty which is heresy. It was Albert Einstein that warned Roosevelt they had to beat the Germans to it, they were already on the path to create it. Kyoto, the religious centre of Japan, is convinced. They want nothing what so ever to do with ThermoNuclear Weapons of Mass Destruction.

Fission and fusion can be allowed in a sufficiently shielded closed controlled environment, but work on weapons of mass destruction in an open system must cease, The Strategic Arms Limitation Treaty is quite pacific. They essentially might have been qualified as extraterrestrials that were working on it and they hated human life!

Even sometimes in History the Entity 'God' can be seen as an Alien Entity, not in the interests of man or beast. Atheists have a good point, that many times this concept and obeyance of 'God' cause too much suffering personally or for humanity in general. Many people would not disagree, that's why there are civil laws, lawyers, and a House of Parliament and a Senate. The division of church and state is an important one historically. It is swings and roundabouts, fearfully, the Church does have to get involved with government from time to time to set the course correction. We need to build the SynGas-Solar hybrid engine based on Artistic concepts of Beauty and Beautiful Scientific principles if we are going to have any hope at all of saving Earth for future human generations. Simultaneously it is onwards and upwards towards Mars and the Stars. We need to support Elon Musk, Jeff Bezos and Richard Branson in their ventures: SpaceX, Blue Origin and Virgin Aerospace respectively, all working with NASA, Boeing, Bombardier,JPL, Caltech, Stanford, MIT and Waterloo and Cambridge University.

When they say it is God and they are trying to get you to do something for them or society without explaining what they want themselves, or without giving any good reason for risking your life without your consent and they won't let you say 'NO' that is definitely NOT GOD the Creator. If they get Violent or attempt to punish you for disobeying, that is definitely NOT GOD either. It was when they conscripted them, it didn't even sound evil. They didn't find out till later that it was evil and it was against their will not 'to live with their family' and against their will to survive and it wanted to risk their life for the gain of others with no actual rewards. Again, the Architect of the Human Brain needs to be involved in the discussion!

Artificial Intelligence-Perhaps the 'NEW GOD'

AI offers a huge $19.9 trillion dollar future over the next 20 years; about the same as the US National Debt. It has the possibility to allow deep learning that doesn't need tedious line code to program and interact with machines and they can provide self-driving cars and trucks, automatic medical diagnosis of medical images, diagnosis of mechanical devices like engines in vehicles and aerospace. Google and Amazon and Facebook and Microsoft are frontrunners in this emerging technology. Little known Canadian Firms hold the key to this enormous market. It can help manage economies, businesses and climate change for governments and scientific institutions and achieve Synergy across all the spheres. This technology can really help solve the Climate crisis and even bring breakthroughs in Medicine on MS and MD and Cancers.

Truly, the new thoughts that AI invokes and creates might move mountains and replace, more pleasurably, any concept of an Ancient God. Remember there still is a Very Ancient God, the Creator of the Universe. It is a work in progress, and we may have to leave for another planet if we screw up Earth too much. A Buddhist might show you how. Even life and death might really be an Illusion, but I am not placing my bets on that. It might all be Maya. The Grand Illusion. Deepak Chopra and Oprah Winfrey might be correct, they are funny and healthy and the Serious people can't take them Seriously. That's why they are Safe. Full of Sound and Peace, signifying Something funny. This is a Religious document, and sometimes you do have to get a bit Serious. Ask Harry Potter about the Dark Lord. I'm siding with Gryfindor and Huff and Puff. We do know it is Slitherin'.

Calcium Carbonate Buffer-Phase I - 2051

Will the Calcium Carbonate Buffer saturate in the future, if so when? And what will be the consequences?

The politicians couldn't understand or respond to this even with all their advisers! pH oscillation in ocean previous results depend on equations from Zeebe and Wolf-Gladrow in book: CO2 in Seawater.

The following article depends entirely on the Diffusion equations for CO2 in water and O2 in water at the air/ocean interface, and the equilibrium at the surface, solved down to depth at varying temperature and pressure. It also, and more importantly, depends on the kinetics equations for the salts and minerals in the ocean like phosphates, nitrates, sulphates, and other known minerals like Calcium Carbonate (aragonite and calcite) and Magnesium (magnesite) and their kinetic balance with CO3(2-),HCO3(-) and H2CO3 in seawater as described in the equations by Dr.Richard Zeebe and Dr. Dieter Wolf-Gladrow in their book: CO2 in Seawater: Equilibria, Kinetics, Isotopes. If those equations are incorrect then my results are incorrect. If they are true, then logically from the software (debugged) the results below are true to the best of anyones knowledge. They can and have been inspected. Also, if the pH equations of Brookhaven National Lab scientist Dr. Ernie Lewis and Doug Wallace are incorrect, then also my results are incorrect. To the best of my knowledge all typos and errors have been removed from the code over a period stretching from 2007 to present. To the best of my knowledge and experience the equations of Zeebe and Wolf-Gladrow and the software and equations of Dr. Ernie

Lewis and Doug Wallace of Brookhaven National Labs are correct and verified. One should bear in mind that 18 g=18 millilitres of H2O contains approx. Avogadros number of molecules of water (plus minerals, etc...) and that is about 6.024 x 10^23 molecules of H2O. According to the work of Dr. K. Gubbins at Cornell, the quantum wave function of 100 molecules of water can not be calculated on a supercomputer because it is too combinatorially explosive. Likewise, an accurate calculation on a quantum computer would take an infinite amount of time, as water moves in time, just like smoke in air. In fact, no accurate calculation of turbulence exists compared to a kayakers mind going down whitewater rapids at the Olympics. Smoke in air is just not repeatable from second to second or millisecond to millisecond for that matter. Likewise, minerals and water structure beyond about 23 molecules of water cannot reasonably be expected to be simulated or predicted with any real to life accuracy. Then there is the contemplation of a lake or a river beyond 18 ml of water, and the contemplation of the whole water cycle, weather, the oceans. Averages into the future can be estimated along with confidence bounds, and these are acceptable for planning, however, actual knowledge of all the trillions and trillions of molecules and their whereabouts from millisecond to millisecond are not predictable. A case in point is flight MH370 from Malaysia to God doesn't know where!!!! if you get my drift. This is my disclaimer for these results. They are averages of what is most likely, but a living ocean and atmosphere with orgone energy throws enough uncertainty into the results that we may never be sure ahead of time. We need to make plans however, and it is wise to take precautions and take the side of caution. I personally am convinced that these results are genuine and we should heed them. But in a complex world with many demands on us like the conflict in Ukraine, Iran, N. Korea,Afghanistan,Pakistan and India,China and the rest of the world, Israel and Palestine, Mexico and US, Energy needs and Oil and Gas or Nuclear (Thorium), and the fact it takes 20 years to get permissions and plan and build a new nuclear reactor.... we may not have enough time to save the oceans. Like James Lovelock said, we might, in order to prevent from getting too depressed over the issue, just enjoy living our little personal lives ; 1 out of 7 billion can't hope to really change it all in time and few would believe him or her anyway because they don't have the background education to understand or the connections necessary, or they might lose too much money or sleep in the process. So C'est la vie. Life goes on one way or the other. Besides, the whole thing may be a secret illusion concocted by aliens, the Elites or both to suck Orgone out of all of us and all of Earth. I wouldn't put it past them. I don't know anyone who has actually taken CO2 measurements of the atmosphere, I've never talked with such a person. I've heard they are using laser spectroscopy to make the measurements at Mauna Kea in Hawaii (the astronomical observatory there) and many other places at NASA, military bases and civilian universities. But I've never seen a TV show about it or heard their tones or found out how it works in detail. It is possible that you can make CO2 measurements with barium hydroxide solution in water in a lab at a school, but it is not very accurate and it is laborious. But i know of no one or no school willing to let their kids do it, or kids willing to do it. The whole thing is based in fear and doomsday messages. I guess we just have to choose to live in spite of the insecurity and realize that it all might be just a bad dream. Babies are still born, and they still drink their mothers milk and they dream, feel secure usually, and grow up and find their way in life one way or the other. The Gig is up. They still kill rats at the hospital in the name of good science. Listen to the music. They either sink or swim. I think we should choose to swim regardless and play some beautiful music and soccer and basketball along the way. The markets are still all stealing and so are the political parties. Politics is often a blood sport. Its not necessary for that to be so with better education and information and connections with good folk. Life is meant to be enjoyed in the presence of good company, and the public news channels are too full of destructive energy and not enough of the orgone to make it pleasant. Check out The Indian Ocean with Simon Reeves..... he ends up in Australia. Pretty good fun. SpaceshipEarth could last another 5 billion years (the expected life of our Sun before it goes Red Giant), if we looked after it properly, NASAs message about leaving and going to another planet and giving up on Earth which they have actively been doing since the 1960s seems altogether doomed aswell unless aliens actually are here and can help. I'm sure some are ok, while others are creating mischief or worse. On that note, the book 'Science,Skeptics and UFOs' written by an retired accredited scientist who has journal papers and patents to his name who comes from farming stock in Mississippi area, whose great grandfather saw

UFOs in the 1890s and his grandfather saw them, his father saw them and he saw them about 40 times in his life. Their womenfolk saw them too. He recounts this personal information in an objective and scientific fashion as possible and his account seems credible. France and the UK and Germany have all declassified their Military UFO sighting and contact research. Britain claims there are aliens (resembling humans very closely) who exist interacting with humans on planet earth right now.

Main Agenda - Phase I - 2051

pH oscillation will impact diversity and life in the ocean perhaps catastrophically unless intervention occurs Fish,crustaceans, corals and marine mammals like cetaceans and all marine animal and plant life may be at risk due to the pH drop that will most likely occur between 2050 and 2053, first the pH will rise by .55 units to 8.95 approximately in 2052 then it will drop to 8.4 the following year then to 7.85 and continue dropping for a while, but not much more until about 2232 when the calcium carbonate buffer breaks at 1329 ppm CO_2. So the calcium carbonate buffer will break or actually find a new equilibrium in the ocean around 493 ppm within a period of 3-4 years, and the calcium carbonate buffer will break in the ocean again around 1329 ppm within a period of 10 years starting around 2232. The calculations take into account the SWS scale of pH which includes an analysis at depth and CO_3^{2-},HCO_3^{-} and H_2CO_3 concentrations and their solubility products as well as phosphates, bromium,boron,chlorine, sodium, Magnesium Carbonate (magnesite) buffer and Calcium carbonate(both aragonite and calcite) buffer and Sulphur dioxide (sulphuric acid) and all other relavant chemical species that exist in the ocean in abundance that are significant (about 98% of the variation is accounted for statistically). An average Temperature change profile with depth is used but temperature changes year over year are ignored in this time period due to the great heat capacity of water, but temperatures are rising causing a die off of phytoplankton in the ocean http://www.theatlantic.com/business/archive/2010/07/phytoplankton-panic/... . A 0.1 degree change in the entire ocean is equivalent to a nuclear bomb going off in terms of raw heat energy. So we ignore the temperature changes as they are mostly negligible on average for the entire ocean. This might be a mistake, a future analysis to take this into account is planned. .

In the interests of time and security and safety these initial results were obtained to get the broad outlines of the chemistry calculated and to release the results. The formulas developed by Brookhaven National Labs at the Carbon Dioxide Information Analysis Centre on the SWS scale for pH and total alkalinity developed by scientists Ernie Lewis and Doug Wallace were used, http://cdiac.ornl.gov/oceans/co2rprt.html and diffusion equations at depth solving for the equilibrium concentration at the surface first. All the formulas in the book CO2 in seawater;Equilibrium, kinetics and isotopes by Zeebe and Wolf-Gladrow were used. http://store.elsevier.com/product.jsp?isbn=9780444509468&pagename=search It is recommend to read the IPCC reports even though their summary only focusses on aragonite, and has better regional data and profiles than the present analysis could muster. https://www.ipcc.ch/publications_and_data/ar4/wg1/en/ch10s10-4-2.html The IPCC report did not contain enough detail to be sure they were reporting the situation accurately or repeatably (they were missing a lot of the story in their public reports and were appealing to their own authority for validity rather than allowing independent verification), which is unscientific. They never admit their ignorance or doubt on any topic. For instance, nobody knows for sure how much aragonite is accessible for buffering from the ocean bottom(floor) across the entire ocean. Estimates have been made by drilling core samples at many locations, and statistical sampling techniques have been used to estimate the total, but nobody knows for sure (they can be pretty certain, but they never even discuss this point). There may be vast deposits of fossil aragonite on the ocean floor that they have missed, or there may be none. My bet is that there are some, given the vastness of the ocean floor, and the fact that they have only just discovered a massive fossil reserve on land near the Burgess shale in the Rockies. http://www.theglobeandmail.com/technology/vast-fossil-bed-found-in-rocki... Also, until the advent of modern sonar techniques (only in 2014 coming to the mainstream science community) to measure salinity, pH,temperature and velocity profiles, only just recently are we getting valid data about the ocean as a whole from these techniques, and there are some errors in measurement that creep in especially if instruments aren't calibrated properly and kept maintained;and there is massive variability in the data, both is space and time. This information has not yet been assimilated by the IPCC team, or at least they have not made this assimilation public, or the data public. I have computed the density and pressure of CO2 concentrations in the atmosphere up to 44 km high (beyond this they are incalculable mostly as they are too small to have an impact). the total is integrated numerically and the results were tabulated. The diffusion equilibrium equations were solved at the surface and at depth to calculate CO2 concentrations at depth for the first 1000 feet of the ocean (333.3 metres), calculated at every .1 metre. The pH values indicate that beginning around 2052 the ocean will exhibit some wild pH fluctuations in the first 1000 feet, and probably beyond at greater depth as well. This relatively large fluctuation from 8.4 baseline up to 8.95 (a difference of .55 pH) and then a dive back down to 7.85 will have a major impact on life in the oceans. fish will feel it as will all marine life. There may be a mass die off, and some evolution. The base of the food chain has already been impacted since 1950, seeing a drop of 40% of phytoplankton population over this time period. http://news.discovery.com/earth/phytoplankton-oceans-food-web.htm The loss of phytoplankton is little understood, but probably related to temperature and pollution and pH changes in the ocean. The loss is probably more attributable historically since 1950 due to temperature and pollution changes in the ocean, as during this time up until the present, pH has stayed relatively stable. We desperately need to find subsets of the marine life and especially krill, zooplankton and phytoplankton and algae that can survive at pH 7.85 (and if they can photosynthesize that would be a bonus) and that are resistant to temperature changes especially. Experiments in the lab at pH 7.85 and increased temperatures with populations of krill and plankton subspecies is necessary to create a population that can survive and thrive under these conditions, and then seed the oceans with this subset at around the right time period.

Timing is everything for this seeding of the Ocean with robust Life. Measurements of ocean pH and temperature need to be made periodically as they are critical to the Timing. Temperature, salinity and velocity profile can be measured in the ocean with new sonar techniques.

Gwynne Dyer: Coral reefs, assisted evolution and geo-engineering

https://www.kamloopsthisweek.com/dyer-coral-reefs-assisted-evolution-geo...

http://spectrum.ieee.org/energy/environment/new-sonar-technology-reveals...http://ieeexplore.ieee.org/xpl/login.jsp?tp=&arnumber=6404791&url=http%3... There are groups of three unmanned drones surveying the ocean as we speak(2014) gathering sonar data, so that triangulation can be done and the equations can be solved for the unknowns.... If the ocean can be seeded at the right time, the base of the food chain will survive, giving optimal chances for survival of aquatic life in general. We need to seed also at a pH of 8.95 and up to this value, and keep seeding on the precipitous dive afterwards. It could be a catastrophic event resulting in the death of the oceans if we don't carefully examine the cause and take preventative and interventionist measures to ensure the survival of aquatic species. It appears the mechanism is that there is a lot less magnesium in the ocean than calcium, and around 2052 the Calcium carbonate buffer suddenly reaches a new equilibrium between 490 to 500 ppm in the atmosphere at sea level (the equilibrium value extending into the surface of the ocean). The phase II - calcium carbonate buffer does not break just yet (thank goodness, and we have some time left to solve the CO2 emission problem after this point due to the ongoing activity of the calcium carbonate buffer). We shouldn't rule out the use of genetic engineering to save the food chain in the oceans, but preferably we should avoid it as it could lead to unintended consequences as natural species have the diversity in their DNA of a long history of environmental changes and the rapid cycling of generations due to pH changes could result in the expression of survival DNA. But there is still a puzzle as to why the zooplankton have not thrived as much or adapted since 1950, perhaps due to pollution in general. Perhaps their DNA has never seen such a pH range and they may be too finely tuned to historic levels prior to the industrial revolution. When the pH reaches 8.95 (which is more basic/alkaline) there will be less CO2 in the ocean, ie. the ocean suck gas CO2 from the atmosphere, increasing pressure and concentration at the depth, causing more CO2 at depth, causing the pH to drop again and become more acidic in a feedback loop. So the rise and fall in pH will oscillate in actual fact and the ppm of the atmosphere will oscillate over relatively short time scales (days and weeks), thus establishing a break in the magnesite buffer. Similarly when the pH is 7.85, there would be more carbonic acid in the ocean, the ocean will emit CO2 into the atmosphere. Once the magnesite buffer is broken, the ppm in the atmosphere directly above and locally will decrease, causing a change in the equilibrium, and the ocean will off gas, increasing the ppm of the atmosphere, and decreasing the CO2 concentration in the ocean, increasing alkalinity and pH in a feedback loop. The ppm and pH will oscillate back and forth in different regions once the feeback loop has started operating and equilibrium would be broken, confirming the break in the magnesite buffer. Unstable ppm and pH levels are dangerous for aquatic species of plants and animals, and the rapid cycling would weather them and perhaps cause an inability to adapt over those time scales involved, resulting in a die off of marine life. The pH could vary by as much as 1.1 pH over the SWS scale over relatively short times like a week or a few days. The feedback loop is practically continuous in time and space. The living ocean is a concept that should not be underestimated.

There are two possibilities that can happen when the calcium carbonate buffer reaches a new equilibrium:

1) Deep ocean CO2 wells up from below and the ocean off gases. When it off gases locally, it causes more diffusion and higher concentrations in the ocean elsewhere.
2) The pH decreases and the acidity increases, drawing CO2 from the atmosphere.

The 2nd isn't as likely as the first because it would take a kind of pump to forcibly remove CO2 from the atmosphere and increase the concentration in the ocean increasing acidity within a two year period by 1 pH level. The equilibrium is the only pump there is and when the buffer breaks, the pump breaks down.

The 1st is more likely, because when the buffer breaks a kind of friction is removed from the depth profile, allowing CO2 from the depths to rush to the surface, overshooting it. The potential is there in the pH of the ocean and as depth increases the CO2 increases, the acidity increases with depth normally. When the equilibrium changes it causes a chain reaction that removes the barriers to upward flow of the dissolved CO2 and the ocean off gases CO2.But then it overshoots and it oscillates back. When the ocean off gases, it increases the atmospheric pressure on the ocean elsewhere causing more diffusion into the ocean there and increasing acidity there, causing an oscillation in pH. This oscillation could weather and harm marine life.

The living ocean is a concept that should not be underestimated. There are waves, pressure and temperature differences, micro-climate hot and cold spots, pH fluctuations and currents and teeming with life (though fish have been disappearing due to the base of the food chain being eroded and overfishing and deep sea trawling). There have been pH values recorded by Al Gore's team of 7.5 in acid spots in the pacific. Perhaps fish feel that and avoid them when they occur due to natural fluctuations. The ocean off gases CO2 from deep ocean currents when they rise to the surface, and highly dense CO2 concentrations exist deep in the ocean in cold waters. Perhaps fish and krill and phytoplankton and zooplankton can already survive at pH 7.5 on average. The averages are still important for the food chain mortality statistics. Research still has to be done to make sure the food chain does not suffer an irreversible collapse if the calcium carbonate buffer suddenly changes to a new equilibrium due to high CO2 concentrations in the top 1000 feet of the ocean. Preventative and Survival measures and intervention might need to be taken to ensure biosphere survival. It is impossible to predict ecosystems and micro-climate accurately or as to timing. The ocean is so variable due to local fluctuations of sunlight, weather and wind and rain and local CO2 concentrations in the ocean and atmosphere, googleplex atoms and molecules dancing about and interacting, chaos and fluctuations and nonlinear dynamical systems theory occurring all the time. But the static theory averages predict that around 490 to 500 ppm of CO2 in the atmosphere, the calcium carbonate buffer could suddenly reach a new equilibrium for the entire ocean within a 3-4 year window. If this happens, the consequences could be drastic. No one can say for sure what will happen. But we should definately be worried and be prepared to take some kind of action geoengineering wise and with assisted evolution to ensure the continued survival of the living ocean and try and prevent us from ever reaching the critical point of 493 ppm CO2 in the atmosphere at sea level. Lets hope that whatever force it takes to save the oceans happens within 35 years in preparation for the event in 2052.

Empirical test of pH decrease, ocean acidification is very simple if you have a chemistry lab near the coast.

Collect 1 litre of seawater from the atlantic, pacific, arctic, antartic or indian ocean in a beaker and seal it and take it to your nearest university chemistry lab.

At the university conduct a controlled experiment with a pH monitor instrument dipped into the beaker and thermometer and pressure measuring instrument(air-barometer) and seal the beaker from the atmosphere except for a CO2 hose going inside the beaker to the air above the seawater.

Slowly increase the concentration and pressure of CO2 above the water and record temperature, pressure in a closed volume along with the time... this process should allow for diffusion of the CO2 into the seawater. Record the changing pH and The CO2 concentration and pressure as the CO2 is slowly over hours increased to 550 ppm at 101.3 kPa equivalent. Using PV=NRT or better yet, van der Walls equations for CO2.

Record the pH drop when it occurs and report the experiment and apparatus online or in a journal. Repeat the experiment 32 times to reduce statistical error to hone the estimate of the crtical empirical concentration of CO2 that causes the sudden achievement of a new equilibrium in the calcium carbonate buffer. Confirm that it is the calcium carbonate buffer that has changed.

Its an advanced calculation of the diffusion equations including gravity, temperature and pressure, after an analysis of the atmosphere to find the ppm at sea level... and assuming equilibrium at the air/sea interface. It is not new physics or chemistry. It uses the ksp solubility product constant equations, derived by scientific experiment, and elucidated in the book

'CO2 in seawater; Equilibria, kinetics, isotopes' by Wolf and Gladrow available from Elsevier press. It is definitely a 'NEW THEORY'.

Here is a thought experiment:

Mammals blood is buffered by sodium carbonate and calcium carbonate. When there is too much CO2 in the blood for the buffer to act upon, the person goes into respiratory acidosis (a well known medical condition). When this happens the medical personnel call it a 'bad day at black rock' and they notify the next of kin.

There is some concentration of CO2 in the Oceans that would also be considered a 'bad day at black rock' and the oceans would be considered to be dying. Politicians and some scientists are refusing to look down that terrible road, like an oil addict who doesn't get to SynGas soon enough or a heroin addict that doesn't get to methodone soon enough or an alcoholic who doesn't get to non-alcoholic beverages soon enough or a smoker who doesn't get to marijuana soon enough! ;-).

CORIOLIS EFFECT: Another interesting phenomenon aside from concentration of ions and gravity is the Coriolis force itself, the force on the oceans molecule that changes with altitude that causes currents to flow along with gravity. Why water in the sink turns onc way in the northern hemisphere and another in the southern hemisphere (clockwise in North, Anti-Clockwise in the South). If the CO2 or carbonate molecules are heavier than water, both gravity and the coriolis effect would help them to sink to depth faster than water would mix. It would cause some water mixing due to friction of molecules. but it also explains the propensity of sea snow to form and fall and redissolve at depth. Modeling this is difficult for the whole ocean as large currents form and some places huge upward currents carrying very dense CO2 concentrations occur and at these places off-gasing occurs.

Here is the data from the program:

pH	Year
8.48011	1970
8.47634	1975
8.47212	1980
8.46762	1985
8.46294	1990
8.45821	1995
8.45357	2000
8.44860	2005
8.44351	2010
8.43922	2015
8.43530	2020
8.43183	2025
8.42936	2030
8.42823	2035
8.42841	2040
8.43061	2045
8.45291	2050
7.93100	2052
8.43359	2053
8.43544	2054
7.92781	2056
7.92597	2058
7.92407	2060
7.91559	2070
7.90541	2080
7.89392	2090
7.88160	2100
7.86922	2110
7.85715	2120
7.84607	2130

7.83623	2140
7.82501	2150
7.82634	2160
7.82214	2170
7.82087	2180
7.82150	2190
7.82306	2200
7.82764	2210
7.82617	2220
7.82505	2230
8.78101	2232
8.62512	2234
8.62365	2236
8.66063	2238
8.64091	2240
8.67210	2250
8.73173	2260
8.81161	2270
8.90857	2280
9.01709	2290
9.12992	2300
9.36401	2320
9.57955	2340
9.80580	2360
10.09740	2380
10.48015	2400

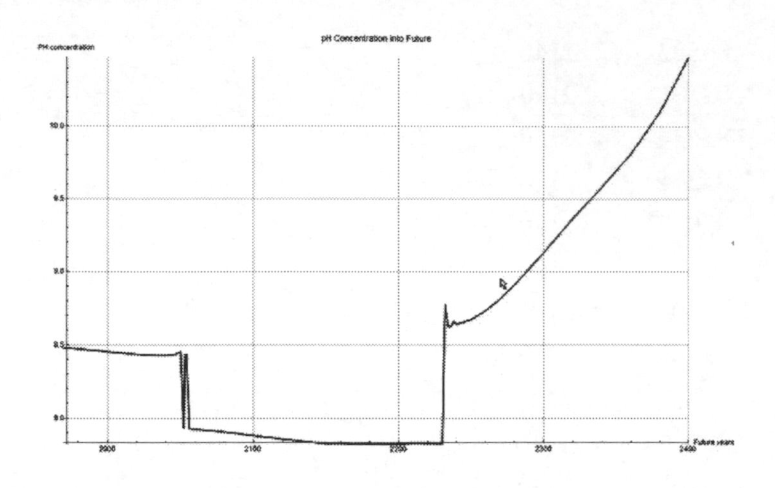

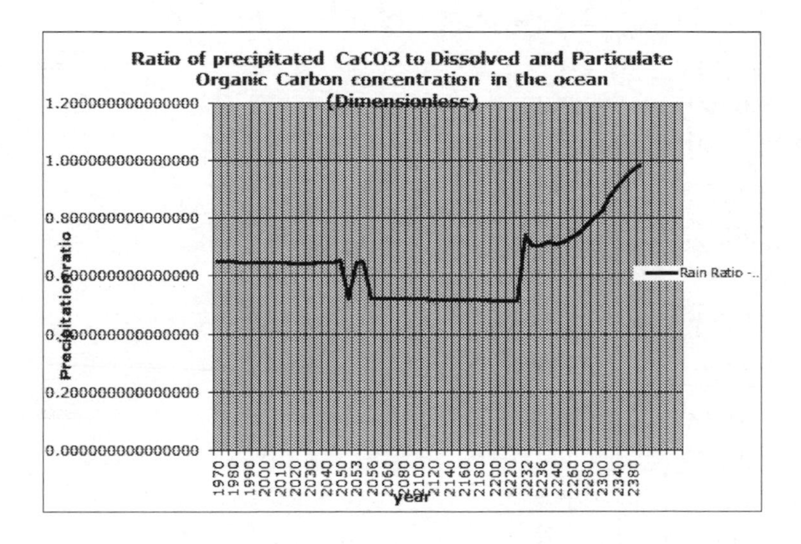

ocean hold

top 1000 metres GT -C	total ocean GT-C	atmosphere GT-C	Ppm (parts Per million)	YEAR
773.28	2870.77	467.92	326.95196	1970.000
775.61	2878.62	476.32	332.82085	1975.000
777.92	2886.36	485.60	339.31052	1980.000
780.25	2894.12	495.77	346.41468	1985.000

782.65	2902.06	506.81	354.12709	1990.000
785.12	2910.18	518.71	362.44147	1995.000
787.67	2918.50	531.46	371.35154	2000.000
790.24	2926.93	545.06	380.85105	2005.000
792.78	2935.29	559.49	390.93372	2010.000
795.39	2943.71	574.74	401.59329	2015.000
798.06	2952.35	590.81	412.82349	2020.000
800.81	2961.21	607.69	424.61804	2025.000
803.61	2970.18	625.37	436.97068	2030.000
806.58	2979.63	643.84	449.87515	2035.000
809.72	2989.59	663.09	463.32517	2040.000
813.04	3000.01	683.11	477.31447	2045.000
818.85	3013.13	703.89	491.83680	2050.000
819.44	3021.25	725.43	506.88586	2055.000
823.13	3032.69	747.71	522.45541	2060.000
826.54	3043.28	770.73	538.53918	2065.000
830.00	3054.01	794.48	555.13088	2070.000
833.48	3064.83	818.94	572.22426	2075.000
837.07	3075.85	844.11	589.81304	2080.000
840.69	3086.97	869.98	607.89097	2085.000
844.35	3098.21	896.55	626.45176	2090.000
848.12	3109.63	923.79	645.48915	2095.000
851.85	3121.09	951.71	664.99688	2100.000
855.81	3132.84	980.29	684.96867	2105.000
859.80	3144.70	1009.53	705.39826	2110.000
863.85	3156.68	1039.42	726.27938	2115.000
867.97	3168.79	1069.94	747.60575	2120.000
872.16	3181.05	1101.09	769.37112	2125.000
876.45	3193.47	1132.85	791.56920	2130.000
880.89	3206.09	1165.23	814.19375	2135.000
885.60	3219.04	1198.21	837.23848	2140.000
891.19	3232.93	1231.79	860.69712	2145.000
940.77	3290.87	1265.94	884.56342	2150.000

894.47	3252.98	1300.67	908.83109	2155.000
900.47	3267.44	1335.97	933.49388	2160.000
905.50	3280.98	1371.82	958.54551	2165.000
910.34	3294.38	1408.22	983.97972	2170.000
915.14	3307.78	1445.16	1009.79023	2175.000
919.95	3321.24	1482.63	1035.97078	2180.000
922.61	3323.68	1559.12	1089.42	2190.000
925.24	3326.07	1637.62	1144.27	2200.000
926.37	3326.97	1718.06	1200.47	2210.000
930.16	3330.53	1800.37	1257.98	2220.000
932.51	3332.65	1884.47	1316.75	2230.000
937.59	3337.50	1970.29	1376.72	2240.000
938.79	3338.47	2057.77	1437.84	2250.000
941.53	3340.98	2146.83	1500.07	2260.000
944.25	3343.47	2237.39	1563.35	2270.000
947.28	3346.26	2329.39	1627.64	2280.000
950.18	3348.94	2422.76	1692.88	2290.000
951.41	3349.93	2517.42	1759.02	2300.000

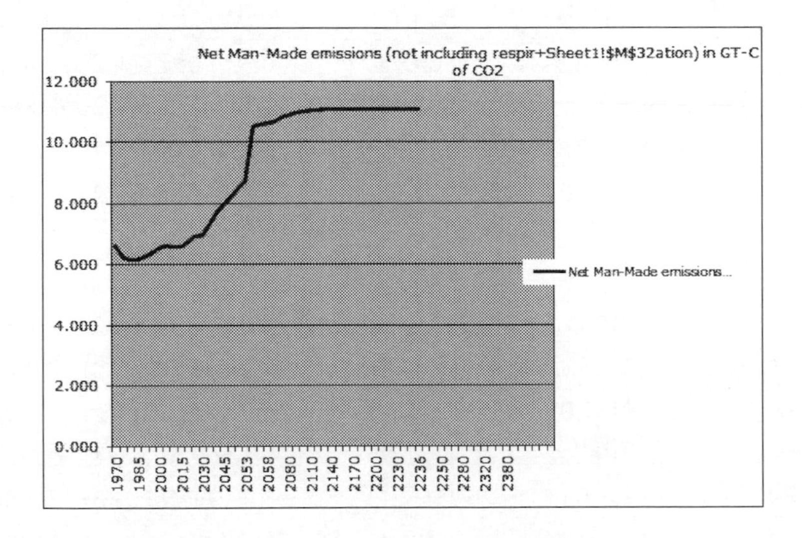

The data for the net Man-Made Emissions comes from a compilation of data from the CIA World Fact Book 2015 and onwards.

The Blue line is the actual data of CO2 parts per million (ppm) in the atmosphere at sea level at Mauna Kea in Hawaii from 1958 to present (Aug 2017). The Green line is a least percent error cubic projection of the data to 2300 if trends continue as they are at present (which is not likely due to the Paris Climate Accord and Cap and Trade on CO2 and societal and technological change in the Future which is very difficult to predict. This is the worst case scenario!) The Red line is the limit of 493 ppm CO2 at which the calcium carbonate buffer reaches suddenly a new equilibrium in the Ocean around 2051 and the Magenta Line is the limit of 1376 ppm CO2 at which the Calcium Carbonate buffer saturates/breaks in the Ocean around 2232.

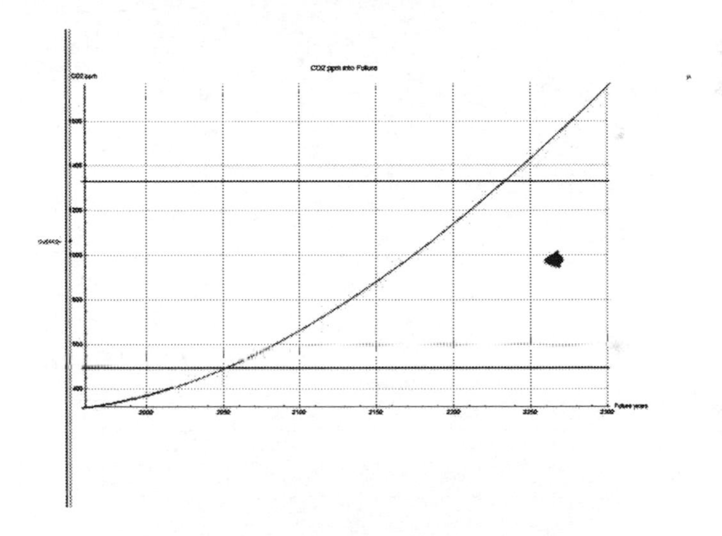

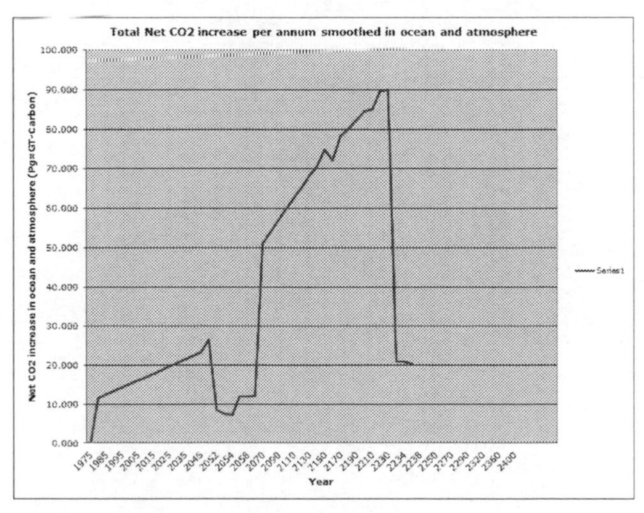

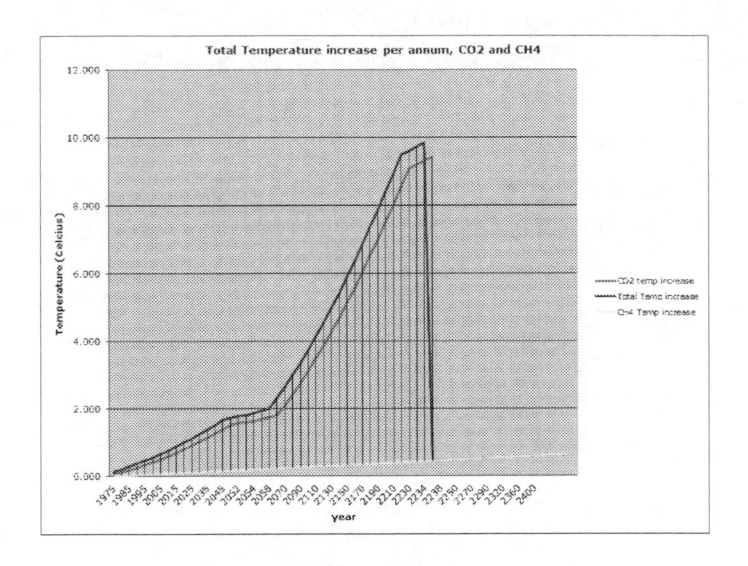
Total Temperature increase per annum, CO2 and CH4
12.000
10.000
8.000
6.000
4.000
2.000
0.000
Temperature (Celcius)
1975
1985
1995
2005
2015
2025
2035
2045
2052
2054
2058
2070
2090
2110
2130
2150
2170
2190
2210
2230
2234
2238
2250
2270
2290
2320
2360
2400
year
CO2 temp increase
Total Temp increase
CH4 Temp increase

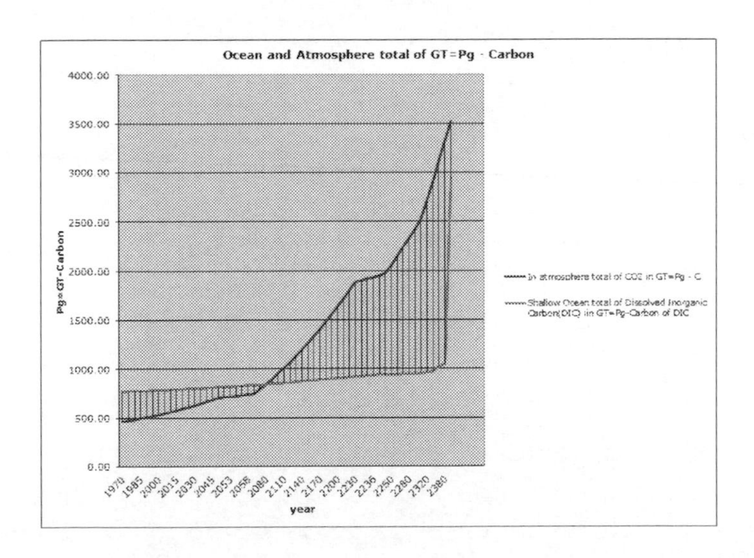
Ocean and Atmosphere total of GT=Pg - Carbon
4000.00
3500.00
3000.00
2500.00
2000.00
1500.00
1000.00
500.00
0.00
Pg=GT-Carbon
1970
1985
2000
2015
2030
2045
2053
2058
2080
2110
2140
2170
2200
2230
2236
2250
2280
2320
2380
year
In atmosphere total of CO2 in GT=Pg - C
Shallow Ocean total of Dissolved Inorganic Carbon(DIC) in GT=Pg-Carbon of DIC

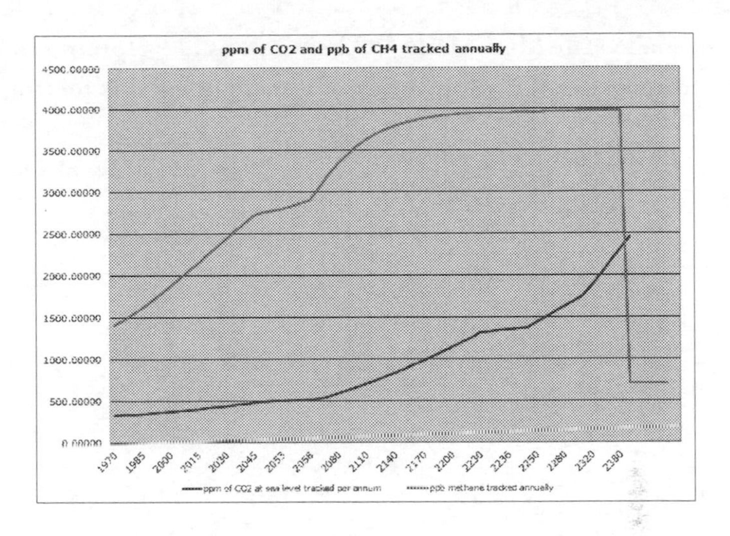

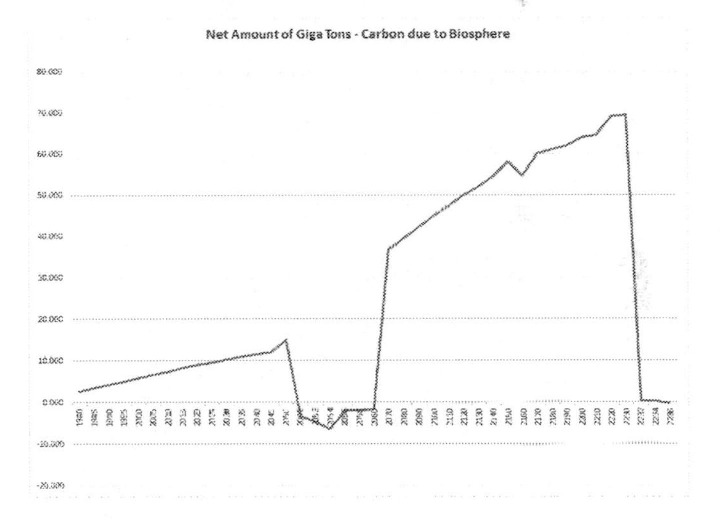

Up until 2052. 2/5 of the problem is man made, after 2052, 1/5 of the problem is man made and the rest **of the CO2 is coming from the Biosphere, not from humanity. We therefore need to suck CO2 out of the atmosphere and out of the Ocean. The highest concentrations are in the Ocean, so our most economical means of preventing the catastrophe are to suck the CO2 out of the Ocean where the deep Ocean currents laden with CO2 come to the surface. We would need rigs, just like oil rigs, whose primary purpose is to separate CO2 from seawater and pipe it to the mainland so that SynGas and Basalt rock**

could be made. The Man-Made CO_2 emissions are a forcing function on the biosphere and the biosphere will respond to this forcing!

https://www.scientificamerican.com/article/pilot-projects-bury-co2-in-basalt/

Calcium Carbonate Buffer Phase II - 2232 - 1329 ppmCO2

By 1339 ppm CO2 in the atmosphere, most likely around 2230 to 2240, the calcium carbonate buffer in the ocean will begin to break and saturate.

The following links are topical for environmental and civilization sustainability. The first one is about the reaching of a new equilibrium of the calcium carbonate buffer in the ocean at 493 ppm of CO2 in the atmosphere and how this will affect the pH of the ocean, and how this in turn will affect the base of the food chain in the ocean, threatening life on the planet in a severe way. The second is about a possible solution to the problem, creating a super-efficient hybrid Syngas- solar engine that emits no CO2, and that could revolutionize energy and transportation globally. The west probably won't do it due to the power of the oil cartels and auto companies and their cash cow status quo. But Europe, China, Russia and India might do it, forcing the west to follow suit.

http://www.thenakedscientists.com/forum/index.php?topic=53181.0
http://www.thenakedscientists.com/forum/index.php?topic=53180.0

When the calcium carbonate buffer begins to break at 1329 ppm CO2 in the atmosphere at sea level, around 2232 annum, the ocean will start to off gas CO2 into the atmosphere in small quantities at first leading to large quantities eventually and stop overall absorbing CO2. Deep ocean currents will well up from the depths and off gas CO2 into the atmosphere more often after this date than previously for a net weather effect.

Around 2232 annum, at around 1329 ppm CO2 in the atmosphere a serious nonlinearity in the equations governing diffusion and CO2 in the atmosphere occurs, and a sharp pH change will occur in the next 10 years affecting all life in the ocean, and possibly all life on the planet because of ecosystem linkages and the interconnectedness of the food chain. Weather will be affected. Unless something drastic occurs in technology before then to prevent the further net emission of CO2 into the atmosphere from the economy, reforestation, and population checks due to developing world initiatives.

The engine mentioned in the Technology section of naked scientists.com that is super fuel efficient and emits only trace waste gases including no CO2 to the atmosphere while emitting trace amounts of carbon monoxide and nitrous oxides, with no net O2 consumption from the atmosphere could really help the situation to save the oceans, but it would be possibly harmful for the oil economy and global stability. It is a challenge of the 21st century to ensure economic stability and achieve a technologically safe earth that isn't harmful to the creatures that live here. It would be a bad metaphor to our future if we went to the stars but our home world died because we couldn't figure out how to save her in time! The knowledge is available, we need political will and international cooperation and trust to solve this problem.

the pH solution that is presented is essentially a mass and charge balance.

The forecast for the ppm CO2 in the atmosphere from Scripps Oceanographic institute is begun with, A longer term forecast is carried further out to 2400, assuming we continue as we are going..... which we won't, as there will be improvements like the blue diesel mentioned by Audi, Tesla and my engine which is zero CO2 emissions......there will be cap and trade and there will also be societal and technological changes by then, including the sonoluminescence lightning resonance engine:

http://www.thenakedscientists.com/forum/index.php?topic=53180.0

http://iopscience.iop.org/book/978-1-6817-4365-3
however, its the most prudent guess right now.

Then the GT-C (mass of $CO2$ carbon) in the atmosphere is solved for based on (Boyles Law)PV=NRT and gravity and density with height.

Then the ppm $CO2$ at sea level is used and an iterative solution of the diffusion equations for equilibrium at the surface is carried out. Solving for equilibrium using gravity and pressure and temperature and salinity gradients with depth and solving the ksp solubility product equations for the concentrations of ions with depth at the balance $CO2$ concentrations and the buffer equations is carried out.... to get the pH with depth. Very small discrete steps with depth were used, and quite a bit of effort went into finding a good and accurate step size.

The average of all the pHs over 1000 metres to get an average pH in shallow water is reported.

ksp solubility product equations are actually behaving like 3D or 4D hyperbolas depending on the formula. Normally, as you start out on one section of the hyperbola, the gradient for the pH is always decreasing as $CO2$ increases, but as it passes the knee of the hyperbola, The concentrations of the buffer equations all start to behave differently in a nonlinear manner.

When this happens, the pH increases instead, and can increase quite dramatically over short periods of time or even decrease dramatically as the case may be.Another way to look at it is that there is a pool of ions fed from 2 or 3 reactions that cascade when the equilibrium is broken by the injection of carbonic acid ions into the solution. This cascade neutralizes the acid by introducing OH- or hydroxyl ions into solution that neutralize or balance the pH. Each buffer has a finite buffering capacity, and it is cumulative. Once the buffering limit capacity is reached, we say the buffer is spent, broken or preferably saturated. After this, the pH can change rapidly.

Once the buffer is broken, the pH does react very differently after the knee of the hyperbola is passed and the buffer equations solution from the ksp solubility products behave very differently too.

The capacity of the buffer to neutralize added acid or added base is compromised at that region of the curve.

When the chemical potential of a dissolved chemical species exceeds that of the same species in solid form then the system will deposit solid(precipitate-sea snow-calcium carbonate) until equilibrium is regained and thereby the system free energy is minimized. Of course that is only one side of the equilibrating system that we are dealing with. This was not modeled in this system, and precipitation is always notoriously difficult to model and involves quantum mechanics and Chaos Theory. Its time to focus on solutions that give us real hope. We know the oceans will have a problem with pH and off gasing, affecting the food chain, life in the ocean, and the global weather. Only solutions that can stop CO2 emissions, while maintaining our standard of living and mobility will work. The engine that emits no CO2 or is CO2 neutral is our best hope. The SynGas-Solar or natural gas -solar or nuclear hybrid engine that is super fuel efficient and only emits trace amounts of carbon monoxide and nitrous oxides and is oxygen neutral offers us real hope. Lets hope saner minds prevail and that reason amounts to something.

The following buffers exist in the ocean: CO2(g) + 2e- -> CO2(2-)(aq) first of all 2 electrons are stolen and there is a charge balance in the ocean so this results in the buffers producing more protons to balance the charge

Ca(2+)+CO2(2-) <-> CaCO2
Mg(2+) + CO2(2-) <-> MgCO2 the magnesium and calcium carbonate buffers steal CO2(2-) and cause sea snow which drops to deeper depths where it redissolves back into Ca(2+) and Mg(2+) and CO2(2-) under pressure.
CO2(2-)+H+ <-> HCO2(-) a proton is stolen, resulting in more hydroxyl radicals than protons.....
HCO2(-)+H+ <-> H2CO2 another proton is stolen resulting in more hydroxyl radicals than protons.
So the net effect of carbonic acid is to first steal 2 electrons from the ocean, resulting in buffers that either reduce the amount of hydroxl radical in the ocean or produce more protons to balance the charge.

next protons are stolen, resulting in buffers that either replace the protons or reduce the hydroxyl radicals or reduce other negative ions in the ocean (most likely) resulting in a net surplus of hydroxyl radicals and a net increase in pH making the ocean more basic. which is what the following data are saying in effect.

H+ + OH- <-> H2O OH steals an electon from H2O and yields a proton and a hydroxyl radical affecting pH or

H+ and OH- combine and form H2O with a net charge change of zero to the charge balance.

Mg(2+)+2HCO2(-_ <-> Mg(HCO2)2

Ca(2+)+2HCO2(-) <-> Ca(HCO2)2 the formation of magnesium and calcium bicarbonate and also its sea snow.

Boron ions, Sulphates (Sulphuric acid and Phosphates (Phosphoric acid).

B(OH)3 + OH- <-> B(OH)4-

B(OH)3 + H2O <-> B(OH)4(-) + H(+)

CO3(2-) + B(OH)3 + H2O <-> B(OH)4- +HCO3(-)

HSO4- <->H+ + SO4(2-)

H2SO4 <-> H+ + HSO4(-)

HF <-> H+ + F-

H3P04 <-> H+ + H2PO4(-)

H2PO4(-) <-> H+ + HPO4(2-)

HPO4(2-) <-> PO4(3-) + H+

And of course

Mg(2+) +2OH(-) <-> Mg(OH)2

Ca(2+) + 2OH(-) <-> Ca(OH)2

and

Ca(2+) + OH(-) + HCO3(-) <-> Ca(OH)(HCO3)

Mg(2+) + OH(-) + HCO3(-) <-> Mg(OH)(HCO3)

Question 1: Where do the 2 electrons come from?, currents in the oceans and lightning. creating storms and charge imbalances.....

Question 2: which species replace the charge in the charge balance? most likely phosphates, sulphates and boron complexes and magnesium carbonate and calcium carbonate.

If the charge is balanced by other species, then there will be an imbalance of H+ and OH- and H2O resutling in a pH imbalance.

It appears from the following data that the net effect of introducing gaseous CO2 in the ocean past 2232 and 1329 ppm CO2 in the atmosphere, is that the ocean becomes more basic, higher pH without bound, also killing the ocean and all marine life, affecting the weather as well and the temperature.

Can we build a SynGas engine that emits no CO2 and is oxygen neutral?

https://www.sciencedirect.com/science/article/pii/S135943111630638X
https://www.sciencedirect.com/science/article/pii/S0263876218300054
https://www.sciencedirect.com/science/article/pii/S1359431116306275
https://arena.gov.au/projects/solar-hybrid-fuels/
http://csp-world.com/news/20130412/00819/new-generation-hybrid-solar-gas-plants-syyngas-and-parabolic-dish
http://www.globalsyngas.org/uploads/downloads/2017-presentations/s13-3-ma.pdf
http://ats.org/the-technion-impact/breakthroughs/environmental-engineering/?gclid=Cj0KCQiAwKvTBRC2ARIsAL0Dgk31P4JzUQl8GeEmoU9bJmd2BWJCm5jrDbEflZdoKl1aDZAhplLLgZ0aAuDzEALw_wcB
https://newatlas.com/solar-gas-hybrid/27073/

It is possible to build a SynGas engine that emits no CO2 and is more fuel efficient than a traditional electric engine like a TESLA egine and consumes no oxygen NET!

SynGas is H2 and CO in the right proportion per volume.... It is a safe fuel if you mix it with enough Argon and Nitrogen to make it less explosive but still combustible enough! Argon to make it less explosive but still combustible enough! It is quite toxic (talk=sic), therefore a closed system with closed valve refuelling is necessary, and driverless cars will make it less likely, hopefully, if they do it right, to have far fewer accidents. Care would need to be taken to insulate the passenger compartment from the fuel system in the event of an accident, and make sure the fuel system is unlikely to leak in the event of an accident. Trying to make it as foolproof as possible however it wouldn't be impossible for a serious leak to occur in a severe accident, which hopefully would be very rare.

You can use Solar power to convert CO2 and H2O, the waste products of burning SynGas with O2 back into the fuel again.

CO2 can be converted separately into CO and 0.5 O2 with a tungsten catalyst.

The H2O separately can be converted to H2 and 0.5 O2.

The extra oxygen can be separated from the CO and H2 with a molecular sieve, or because of point of origin like with H2O, because you need two electrodes, often made of platinum, but steel will suffice.

You need to mix the H2O with potassium (K) to form an electrolyte that will allow you to electrolyze it more efficiently.

When you remix the H2 and CO you have the fuel again.

The solar power can be used to electrolyze the H2O, and a heat pump with heat from the waste products of the engine can be used to use a higher temperature to catalyze the CO2 into CO and O2. These can be separated with a molecular sieve. The oxygen should be released to the atmosphere by premixing it with enough air and exhausted to the atmosphere. This engine is oxygen neutral and does not NET consume oxygen.

It is possible to build the engine without a molecular sieve, if you mix the CO and .5O2 with Argon and air to the right proportion before injecting it into the combustion chamber to mix with the H2. The H2 is naturally separated from the O2 because H2 comes physically from the negative electrode and O2 comes from the positive electrode thus they are pysically separated in space and can be kept separated, it is best to offgas the O2 from the electrolysis by premixing it with air or keep it to mix in with the

CO and O2 and air (nitrogen and argon and O2) prior to injecting into the cylinders.

H2 does make metal brittle and corrodes it over time, so the H2 pump and fuel lines and H2 injectors will probably need to be replaced everty 3-10 oil changes.

The engine is practical and the O2 never has to exist with the fuel at the same time, unlike a methane cycle and so is much safer than a methane recycling engine.

It might be possible to get from 3000 to 8000 km on a single tank of 55Litres of SynGas.

We wouldn't need a war for Oil anymore!

It might make a big revolution for Aviation, as the range would be considerably extended and there would be enough oxygen to make the cycle work even at high altitudes, so it might be possible for them to fly above the weather.

Check out:

https://today.uic.edu/breakthrough-solar-cell-captures-co2-and-sunlight-produces-burnable-fuel

https://www.scientificamerican.com/article/can-carbon-dioxide-replace-steam-to-generate-power/

https://www.theguardian.com/environment/2010/may/06/co2-green-fuel-car

http://newatlas.com/molecule-co2-carbon-neutral-fuel/48390/

Very important: The theoretical perfect efficiency of an engine is (T1-T2)/T1 where T1 is the temperature of the hotter reservoir and T2 is the temperature of the cooler reservoir. This formula was probably discovered by Lord Kelvin, but it was first described and directly followed from the work of Sadi Carnot:

https://en.wikipedia.org/wiki/Nicolas_L%C3%A9onard_Sadi_Carnot

https://en.wikipedia.org/wiki/William_Thomson,_1st_Baron_Kelvin

A perpetual motion machine is actually impossible, as entropy and friction and losses exist, but we may be able to get almost infinitely close to it eventually!

https://en.wikipedia.org/wiki/Perpetual_motion

Time Crystals bypass the laws of themodynmaics themselves as they are not a closed system.

If we go to an open system, with access to energy from the Universe, it might be possible to have what seems like a perpetual motion machine to the uninitiated. The energy has to come from somewhere, like the magnetic field of the earth. If we draw energy from the magnetic field of the earth, and everybody does it, it might eventually deplete the magnetic field of the earth and stop it from protecting us from solar system Cosmic Rays. Causing cancers and illness. It might even not be safe to be around such a machine as strong fields can cause illness.

The engine described above is NOT a perpetual motion machine and needs refilling eventually, however it is more efficient than a traditional electric engine that does not draw energy from the magnetic field like the TESLA engine. This engine is not a closed system, and does draw extra energy from solar power to do the electrolysis, and a 1 square metre solar panel is sufficient to accomplish this on a 4 passenger motor vehicle. The heat pump can be powered by the engine and a battery and an electric alternator and the battery can be recharged by the alternator.

Forecasting: Turbulence - The Stock Market - Earthquake Probability

Forecasting Turbulence and the Stock Market: The idea here is to show a way with VectorVest and this new software in Matlab and Mathematica to generate funds from the stock market on an ongoing basis, with a relatively small initial investment that grows with time and allows us to fund the design and development of the engine so that the fear of risk and the large initial outlay can be alleviated. This reduces the activation energy for the investment and makes it far more likely to occur.

Density Functional American Option pricing with Bayesian Monte Carlo Path Int & MUSIC w/ Kelly Crit

A probability distribution is calculated from a past stock chart and the fat tails are estimated. The distribution, unique to every stock chart, is used to generate an arbitrary number of paths into the future (3000 by default) and 50% and 80% and 98% confidence intervals are calculated and displayed. A computation of the most probable call to put ratio is calculated and used to generate a Most Probable Win percentage per day. The final absolute probability of winning at the 98% level is used to calculate the Kelly Criterion multiplied by 0.92 in the Risk Analysis section of the figure. This simulation differs significantly from Black Scholes and is more accurate. Every stock has its own probability distribution in the past. Care should be taken to note that market conditions can change abruptly in the future and distributions can be significantly altered over a short time frame by global market events, Intra-day distributions can be remarkably different from end of day distributions, so the results may vary from real world trading but it is probably the most accurate look ahead in stable markets that we are going to get with a mathematical approach. Care and attention to liquidity, volume and volatility should be observed. Stock Forecasting and American Option pricing compared to Black- Scholes(BS) using Bayesian Markov Monte Carlo Simulation and Wavelet or Fourier or Neural Network Extrapolation with Indicators..http://library.wolfram.com/infocenter/MathSource/9086/ for more information! If you use the alternate program as well, Bayesian Markov Stochastic Monte Carlo Valuation of Integrated Price Volume Action with Kelly Criterion, the kelly values from that program should be more accurate as an assessment of volume trends are also incorporated and we all know how important liquidity is. An expert using both programs will be a better judge just before an earnings announcement. A Fourier extrapolation is also carried out, but is only valid if it is supported in the range by the Statistical look ahead! The extrapolation is important for assessment of trends and cycles!

https://www.mathworks.com/matlabcentral/fileexchange/56352-density-functional-american-option-pricing-with-bayesian-monte-carlo-path-int—music-w—kelly-crit

Density Functional Bayesian Monte Carlo Valuation of Price Volume Action w/MUSIC & Kelly Criterion

This program is very similar to Bayesian Markov Stochastic Monte Carlo American Option Pricing with Kelly Criterion. It uses special code to combine price and volume into an action over time which is integrated to see the temporal effect of price and volume. The value displayed by the call and put sections and black scholes is not a real dollar amount, but a value of integrated price volume action. To give some idea about how price and volume will affect value over time. In conjunction with the other program, an expert can judge whether an investment is worth it or not and what the risk is and what the probability of winning within a certain amount of time is. The kelly criterion is still valid and one should use the kelly criterions from this program and not the dollar program. Care should be taken over whether to wager with the Fourier extrapolation trend or whether to go with the statistics. Only an expert can judge if they are saying different things. If BOTH the statistics and Fourier trend are on the same side of the ledger, you can bet that your wager is safer still. A MUltiple SIgnal Classification is also carried out (MUSIC) which allows Fourier Extrapolation is also carried out.

https://www.mathworks.com/matlabcentral/fileexchange/56446-density-functional-bayesian-monte-carlo-valuation-of-price-volume-action-w-music—kelly-critrion

Density Functional Stock Forecasting and American Option pricing compared to Black-Scholes(BS) using Bayesian Markov Monte Carlo Simulation and Wavelet or Fourier or Neural Network Extrapolation with Indicators.

A probability Density Functional Transform (DFT) is created from the past years stock price chart and the day to day ratios of price magnitude to allow us to estimate accurately the Markov Transition probabilities from one day to the next. The distribution of the number of occurrences or frequencies of the ratios creates the Density Functional. Then we use these to generate 3000 to 4000 paths into the future, with probabilities and ratios for every single day leading each day to the next price, and a most probable path with confidence limits is calculated. The 25,75, and 10,90 and 2,98 % confidence limit pair paths are calculated and displayed. The most probable path is displayed after path integration and with beautiful graphics. A Fourier extrapolation is carried out with Multiple Signal Classification technique (MUSIC) and low frequencies are estimated from a high resolution Fast Fourier Transform(FFT). The magnitude and phase of these root frequencies is calculated with the Discrete Fourier Transform (DFT) from the data and is then extrapolated into the future. This is relatively fast, efficient and accurate as only a small number of root frequencies are really needed to get the form of the data to a high level of approximation. Also for research interest, A novel neural network that uses no backpropagation but a functional insight and random perturbations that is faster and more accurate than backpropagation can be used as welll. A Wavelet extrapolation is calculated from computing the wavelets of the data, and then at each level, doing a Fourier extrapolation of the wavelet coefficients. This is thw Wavelet Functional Theory(WFT). WFT is less accurate than Fourier Extrapolation and more computationally complicated. Usually only the Fourier extrapolation and the Path integral Markov Monte Carlo Simulation is performed and this takes about 2 minutes and 20 seconds for 1 years data and a 6 month lookahead. Another breakthrough was the conception of the Integrated Price-Volume Action, which is a novel indicator chart called the Accumulation that shows if price and volume have been increasing together over the past year or have been declining or have been moving at odds with each other. Forecasting this chart with Markov Monte Carlo Path Integration and Fourier Extrapolation is also done, and it is a more realiable indicator of future performance than the analysis on the price chart alone! These methods allow us to accurately predict the true value of profit from a Call or a Put, and give confidence limits as well. The most probable statistical win percentage per day is calculated from the most probable call-put ratio per day. All the other normal traditional indicators are calculated as well and 4 legs can be calculated at once to try different call-put strategies, like strangles and iron condors. With this tool, you might find yourself relying on a bare call or put more often as you will know the probability of success. The Kelly criterion is calculated, and this automatically gives you the right amount of your purse to bet on a given strategy based on the odds of winning! Other work that i have worked on has to do with saving the oceans and creating an engine or power plant that emits no CO_2 with SynGas, check out Ocean Acidity Climate Shock article at: http://www.thenakedscientists.com/forum/index.php?topic=6013 2.msg484012#msg484012 At the Science Forum at the University of Cambridge in England! See also for reference, past work: https://www. greenparty.c a/sites/default/files/picketfence.pdf See also implementation in MATLAB for comparison: https://www.mathworks.com/matlabcentral/ profile/authors/1738497-chondrally

All the relevant indicators are displayed for a timely and a long or short term investment decision based on the probability of success both in terms of the most probable path into the future and in terms of all paths into the future. In that sense this is a path integration technique that employs Bayesian and Markov thinking with a multi path Monte Carlo simulation. Alternatively Fourier Extrapolation or Wavelet Extrapolation or Neural Network training and extrapolation are also employed. Novelly, the Nets use no backpropagation and when they work they find the phase, amplitude and frequency of the main cycles more accurately than backpropagation (they find a better global minimum solution). They work about 90% of the time, and need to be rerun until the output makes sense. You will generally get slightly varying local minima each time so expect the answer to be a bit different each time from the neural nets. The probable answer and confidence limits are always unequivocal for the monte carlo simulation (the blue path and red confidence limits). The Wavelet and Fourier extropolations are also unequivocal. To Train the nets set optionflag=1 and the program will save the weights in OutOptions2.txt in the directory of your choice. To run the program once the weights are trained for different spot prices and expiry dates or different inflation or interest rates or dividend rates, set optionflag=0. If you do not wish to use the neural nets set optionflag= negative one=-1 for just the Bayesian Markov Monte Carlo Simulation or set optionflag =2 for the path Simulation and default Fourier Extrapolation wihich is the fastest and most accurate. If optionflag=3, a wavelet extrapolation and discretewavelettransform using Biorthogonalsplinewavelet[4,2] as the default. The simulation is also carried out for comparison purposes and full statistical analysis of all paths. If you wish to see the progress of the evaluation set debugflag=1. Alternatively set debugflag=0 which is the default and runs quite faster. Without the neural nets the program evaluates in 2-3 minutes depending on the length of time to expiration of the option, the lookback period and the time to expiry of the option. With the neural nets it runs in under 11 minutes on a modern PC. 3GHz intel 7 processor with 4 cores. If catastropheflag =1, a multi hour computation is carried out on an ETF or index like the "SP500" to determine the daily probability of level crossing into the future. The levels are the ratio of the next day to the previous day price. and the images are the probability of crossing the level per day with ratios of.4,.5,.6,.7,.8,.9 and also 1.1,1.2,1.3,1.4,1.5,1.6 this can take up to a 9 hours to compute. Normally catastropheflag must be set to zero for normal stock analysis. It can be run on a supercomputer to really speed things up.The simulation can be parallized if that is wished but would take a bit of work. the key routines involved would be generateMagPhase, calculateFutureValues and CalculateConfLevels and NeuralNet1. Because of the different and varied results offered by the neural nets and that they fail some of the time, no evaluation of the option price is based on them. The unequivocal answers of the Monte Carlo simulation and the confidence limits of between 1000 and 4096 or more paths into the future are calculated and the results are reliable. Also so are the Wavelet and Fourier extrapolation (optionflag =3 and optionflag=2 respectively). and option values are calculated on the wavelet and fourier results aswell. These can be compared with the Black-Scholes calculation shown at the bottom of each output panel. Periodicities in the time series are not well predicted by the Monte Carlo Simulation, however theWavelet Extrapolation and Fourier Extrapolation and Neural Nets do a good job of predicting the lows and the highs. The nets are reliable enough to be included for the expert to evaluate and the Wavelet and Fourier results are yielded everytime. The expert can make a rational judgment about the nature of the periodicities (phase, amplitude and frequency) of the cycles in the time series. To get Wavelet extrapolation, Fourier Extrapolation Neural Nets and Bayesian Markov Path integration use OptionsHospitality3Alpha11.nb you will need the Wavelet module available in Mathematica for this. To use the program without wavelet module, but with fourier extrapolation and neural nets aswell as the Bayesian Markov Monte Carlo Path Integration, use OptionsHospitality3alpha9.nb OTHER AUTHORS: Michael Kelly helped debug the intial code. The idea and concept and design are the original authors. . with input from Sornette et al were used in evaluating the extreme values of a distribution with fat tails.

http://library.wolfram.com/infocenter/MathSource/9086/

This software could be a breakthrough in Turbulence forecasting and could be applied to airplanes, with sonar lookahead.

Earthquake Forecasting:

Earthquake probability Distribution code

(* COPYRIGHT 2012,2013,2014,2015,2016,2017
CHONDRALLY/PSYREIGHE, ASA, AGU, IEEE, AAAS, ACS, ACM
ALL RIGHTS RESERVED)
(* Forecasting probability, magnitude and time of occurrence \
software using time series and Density Functional Bayesian Martingale \
Monte Carlo Path Integration and Fourer Extrapolation.
This software can Forecast the likely times and magnitudes of \
earthquakes or any events with a magnitude and a time stamp. It can \
be used to forecast terrorist strikes if a past time series of \
magnitudes and timestamps is known for certain regions or cells. It \
can be used to forecast losses or wins in military campaigns. it can \
be used to forecast emergency calls if they are ranked in magnitude \
with a time stamp. it can be used to forecast suicide calls or \
distress calls. it can be used to forecast crimes and murders and \
robberies if a magnitude (category) is assigned with a time stamp. \
It can help solve serial killings and serial crimes and drug deals. \
it can be used in another form to forecast the stock market or stock \
market crashes or surges. it can be used to forecast extreme events \
of all kinds like hurricanes and tsunamis and volcanic cruptions. It \
can be used to forecast traffic jams and traffic accidents or any \
rare events like hijackings or airplane accidents or events like 911 \
or the occurrence of declarations of war between nations or more \
importantly the occurrence of trade deals or peace treaties. it can \
be used to forecast the outbreak of riots or rebellions or the \
occurrence of music concerts and the generation of new pop - stars. \
It can be used in epidemiology to forecast the outbreak of diseases \
in many regions of the world and outbreaks of manias, panics and \
crashes. I was inspired by Isaac Asimov's book 'The Foundation Trilogy' especially by the character: 'Hari Seldon – The PsychoHistorian'
*)

Fluid Companionship - A reminder about the limits of Mathematics:

Each drop's splash seen from the window up on high as the edge of the wind wafts past in wave fronts of rain.

Like wind in the long grass, flowing, alternating shades of green.

Like quicksand collapsing to consume itself, or so it seems.

Like water coursing, etching patterns never to be seen.

Like a plume from an undersea geyser-vent, hidden from view.

Like drifting tendrils of snow on a field, poetry in motion.

Like a lava avalanche from a volcano, beautiful and deadly, scorching.

Like dust devil's, coalescing, dancing, darting, dispersing.

Like a puff of smoke into a beam of light.

Like a violet, orange, red and pink Sunrise, full of motion and promise!

Like a Commando's mind when he takes the life of a supposed enemy.

Like a sailor, out on the water, jibing and tacking.

Like a hangglider, floating high on thermals.

Like a Scuba Diver down at 105 feet, getting only a glimpse.

Like a Skier's mind, racing down the moguls.

Like a Kayaker's mind going down a series of whitewater rapids.

Like a Fireman entering a burning building, pulling a little girl to safety.

Like a warm shower on a cold winter's day, full of billoughy steam.

Like a cat purring on your lap in the evening.

Like fire, shifting, shining, reaching for the sky.

The High and the Low colliding, a lightning storm created.

From the window, warm and protected:

Like Dreams…Like no Mathematics can fully describe!

DoomsDay Clock - 2 seconds to Midnight

http://www.cbc.ca/news/technology/doomsday-clock-2018-1.4502382 It seems that the Doomsday clock is actually much closer to midnight than the Scientists who program the doomsday clock are letting on, if we take into account the oil, gas and nuclear and defence industries as well as the current state of the Oceans and Ocean and atmosphere temperatures and pH and pollution worldwide aswell as famine, drought and disease and extreme weather events and earthquakes and volcanos. We are all going to win a Darwin Award by 2051 if We don't do the correct thing about it. http://www.darwinawards.com/darwin/ Remember the Warning of the Story of Ceaucescu! https://en.wikipedia.org/wiki/Nicolae Caeucescu
Monty Python – Life of Brian and The Secret Policeman's other ball!

Appendix A - Average number of Earthquakes for Various Countries for 10,20 and 30 years and time till at least one quake of each magnitude.

Japan

magnitude cutoff = 5., years = 0.0060782, Probability= 1.7442*10-6 per quake or 1.51211*10-6 per day, count = 1.0302, sd = 0.856922, mean number of quakes in time period is: 0.641533, Number of days in period = 2.22001

magnitude cutoff = 5., years = 10., Probability= 1.44319*10-6 per quake or 6.60575*10-7 per day, count = 519.05, sd = 163.281, mean number of quakes in time period is: 1253.83, Number of days in period = 3652.42

magnitude cutoff = 5., years = 20., Probability= 1.45356*10-6 per quake or 6.55754*10-7 per day, count = 1028.07, sd = 225.086, mean number of quakes in time period is: 2471.61, Number of days in period = 7304.84

magnitude cutoff = 5., years = 30., Probability= 1.44113*10-6 per quake or 6.47719*10-7 per day, count = 1521.93, sd = 287.442, mean number of quakes in time period is: 3693.57, Number of days in period = 10957.3

magnitude cutoff = 5.5, years = 0.00824855, Probability= 1.2463*10-6 per quake or 1.05116*10-6 per day, count = 1.0604, sd = 0.903376, mean number of quakes in time period is: 0.847, Number of days in period = 3.01272

magnitude cutoff = 5.5, years = 10., Probability= 8.55162*10-7 per quake or 3.91423*10-7 per day, count = 341.551, sd = 117.962, mean number of quakes in time period is: 1253.83, Number of days in period = 3652.42

magnitude cutoff = 5.5, years = 20., Probability= 8.55461*10-7 per quake or 3.85929*10-7 per day, count = 672.222, sd = 160.519, mean number of quakes in time period is: 2471.61, Number of days in period = 7304.84

magnitude cutoff = 5.5, years = 30., Probability= 8.45805*10-7 per quake or 3.80148*10-7 per day, count = 992.598, sd = 202.782, mean number of quakes in time period is: 3693.57, Number of days in period = 10957.3

magnitude cutoff = 6., years = 0.0120464, Probability= 7.9613*10-7 per quake or 6.46189*10-7 per day, count = 1.035, sd = 0.92168, mean number of quakes in time period is: 1.1904, Number of days in period = 4.39986

magnitude cutoff = 6., years = 10., Probability= 4.63133*10-7 per quake or 2.11984*10-7 per day, count = 208.3, sd = 76.0192, mean number of quakes in time period is: 1253.83, Number of days in period = 3652.42

magnitude cutoff = 6., years = 20., Probability= 4.58509*10-7 per quake or 2.0685*10-7 per day, count = 406.31, sd = 102.645, mean number of quakes in time period is: 2471.61, Number of days in period = 7304.84

magnitude cutoff = 6., years = 30., Probability= 4.51742*10-7 per quake or 2.03036*10-7 per day, count = 598.145, sd = 128.835, mean number of quakes in time period is: 3693.57, Number of days in period = 10957.3

magnitude cutoff = 6.5, years = 0.0195717, Probability= 5.07479*10-7 per quake or 3.89278*10-7 per day, count = 1.0166, sd = 0.969414, mean number of quakes in time period is: 1.8278, Number of days in period = 7.14839

magnitude cutoff = 6.5, years = 10., Probability= 2.51166*10-7 per quake or 1.14963*10-7 per day, count = 113.234, sd = 42.7155, mean number of quakes in time period is: 1253.83, Number of days in period = 3652.42

magnitude cutoff = 6.5, years = 20., Probability= 2.46348*10-7 per quake or 1.11136*10-7 per day, count = 218.746, sd = 57.2696, mean number of quakes in time period is: 2471.61, Number of days in period = 7304.84

magnitude cutoff = 6.5, years = 30., Probability= 2.42086*10-7 per quake or 1.08806*10-7 per day, count = 321.161, sd = 71.4552, mean number of quakes in time period is: 3693.57, Number of days in period = 10957.3

magnitude cutoff = 7., years = 0.0348961, Probability= 3.00222*10-7 per quake or 2.1828*10-7 per day, count = 1.0294, sd = 1.05266, mean number of quakes in time period is: 3.08893, Number of days in period = 12.7455

magnitude cutoff = 7., years = 10., Probability= 1.29309*10-7 per quake or 5.91869*10-8 per day, count = 59.3048, sd = 22.7366, mean number of quakes in time period is: 1253.83, Number of days in period = 3652.42

magnitude cutoff = 7., years = 20., Probability= 1.26*10-7 per quake or 5.68433*10-8 per day, count = 113.8, sd = 30.485, mean number of quakes in time period is: 2471.61, Number of days in period = 7304.84

magnitude cutoff = 7., years = 30., Probability= 1.23574*10-7 per quake or 5.55407*10-8 per day, count = 166.735, sd = 37.8672, mean number of quakes in time period is: 3693.57, Number of days in period = 10957.3

magnitude cutoff = 7.5, years = 0.130559, Probability= 9.87637*10-8 per quake or 6.02525*10-8 per day, count = 1.1212, sd = 1.10939, mean number of quakes in time period is: 9.69713, Number of days in period = 47.6856

magnitude cutoff = 7.5, years = 10., Probability= 4.23031*10-8 per quake or 1.93629*10-8 per day, count = 20.6904, sd = 8.10272, mean number of quakes in time period is: 1253.83, Number of days in period = 3652.42

magnitude cutoff = 7.5, years = 20., Probability= 4.07312*10-8 per quake or 1.83753*10-8 per day, count = 39.2638, sd = 10.8333, mean number of quakes in time period is: 2471.61, Number of days in period = 7304.84

magnitude cutoff = 7.5, years = 30., Probability= 3.98033*10-8 per quake or 1.78897*10-8 per day, count = 57.336, sd = 13.3958, mean number of quakes in time period is: 3693.57, Number of days in period = 10957.3

magnitude cutoff = 8., years = 6.99321, Probability= 2.74019*10-9 per quake or 9.47766*10-10 per day, count = 0.9428, sd = 0.760083, mean number of quakes in time period is: 294.48, Number of days in period = 2554.21

magnitude cutoff = 8., years = 10., Probability= 2.51323*10-9 per quake or 1.15035*10-9 per day, count = 1.22725, sd = 0.666565, mean number of quakes in time period is: 1253.83, Number of days in period = 3652.42

magnitude cutoff = 8., years = 20., Probability= 2.20265*10-9 per quake or 9.93696*10-10 per day, count = 2.12025, sd = 0.880572, mean number of quakes in time period is: 2471.61, Number of days in period = 7304.84

magnitude cutoff = 8., years = 30., Probability= 2.08597*10-9 per quake or 9.37542*10-10 per day, count = 3.00065, sd = 1.06559, mean number of quakes in time period is: 3693.57, Number of days in period = 10957.3

Country = Russia

magnitude cutoff = 5., years = 0.00186865, Probability= 2.5366*10^-6 per quake or 5.19207*10^-6 per day, count = 0.9985, sd = 0.912767, mean number of quakes in time period is: 0.155222, Number of days in period = 0.682508

magnitude cutoff = 5., years = 10., Probability= 2.99883*10^-6 per quake or 2.01279*10^-6 per day, count = 1535.4, sd = 306.957, mean number of quakes in time period is: 2206.33, Number of days in period = 3652.42

magnitude cutoff = 5., years = 20., Probability= 3.03846*10^-6 per quake or 2.28295*10^-6 per day, count = 3491.88, sd = 721.309, mean number of quakes in time period is: 4939.65, Number of days in period = 7304.84

magnitude cutoff = 5., years = 30., Probability= 3.04936*10^-6 per quake or 2.53076*10^-6 per day, count = 5810.84, sd = 1219.12, mean number of quakes in time period is: 8184.41, Number of days in period = 10957.3

magnitude cutoff = 5.5, years = 0.00233525, Probability= 1.76559*10^-6 per quake or 3.34516*10^-6 per day, count = 0.9945, sd = 0.91098, mean number of quakes in time period is: 0.179556, Number of days in period = 0.852931

magnitude cutoff = 5.5, years = 10., Probability= 1.73651*10^-6 per quake or 1.16553*10^-6 per day, count = 1198.99, sd = 254.746, mean number of quakes in time period is: 2206.33, Number of days in period = 3652.42

magnitude cutoff = 5.5, years = 20., Probability= 1.76932*10^-6 per quake or 1.32938*10^-6 per day, count = 2734.17, sd = 576.09, mean number of quakes in time period is: 4939.65, Number of days in period = 7304.84

magnitude cutoff = 5.5, years = 30., Probability= 1.77825*10^-6 per quake or 1.47583*10^-6 per day, count = 4553.17, sd = 966.725, mean number of quakes in time period is: 8184.41, Number of days in period = 10957.3

magnitude cutoff = 6., years = 0.00310678, Probability= 1.22512*10^-6 per quake or 2.11128*10^-6 per day, count = 0.9775, sd = 0.897704, mean number of quakes in time period is: 0.217278, Number of days in period = 1.13473

magnitude cutoff = 6., years = 10., Probability= 1.06769*10^-6 per quake or 7.16624*10^-7 per day, count = 890.862, sd = 198.484, mean number of quakes in time period is: 2206.33, Number of days in period = 3652.42

magnitude cutoff = 6., years = 20., Probability= 1.08888*10^-6 per quake or 8.18127*10^-7 per day, count = 2032.13, sd = 436.191, mean number of quakes in time period is: 4939.65, Number of days in period = 7304.84

magnitude cutoff = 6., years = 30., Probability= 1.09475*10^-6 per quake or 9.08572*10^-7 per day, count = 3384.64, sd = 726.487, mean number of quakes in time period is: 8184.41, Number of days in period = 10957.3

magnitude cutoff = 6.5, years = 0.00458933, Probability= 8.58061*10^-7 per quake or 1.26671*10^-6 per day, count = 0.9475, sd = 0.87437, mean number of quakes in time period is: 0.274944, Number of days in period = 1.67622

magnitude cutoff = 6.5, years = 10., Probability= 6.54016*10^-7 per quake or 4.3897*10^-7 per day, count = 614.914, sd = 142.483, mean number of quakes in time period is: 2206.33, Number of days in period = 3652.42

magnitude cutoff = 6.5, years = 20., Probability= 6.65099*10^-7 per quake or 4.99722*10^-7 per day, count = 1399.07, sd = 304.53, mean number of quakes in time period is: 4939.65, Number of days in period = 7304.84

magnitude cutoff = 6.5, years = 30., Probability= 6.68494*10^-7 per quake or 5.54805*10^-7 per day, count = 2329.55, sd = 504.441, mean number of quakes in time period is: 8184.41, Number of days in period = 10957.3

magnitude cutoff = 7., years = 0.00759372, Probability= 6.05815*10^-7 per quake or 7.76068*10^-7 per day, count = 0.9975, sd = 0.925974, mean number of quakes in time period is: 0.394778, Number of days in period = 2.77355

magnitude cutoff = 7., years = 10., Probability= 3.73107*10^-7 per quake or 2.50426*10^-7 per day, count = 375.321, sd = 89.4567, mean number of quakes in time period is: 2206.33, Number of days in period = 3652.42

magnitude cutoff = 7., years = 20., Probability= 3.77871*10^-7 per quake or 2.83914*10^-7 per day, count = 850.754, sd = 186.978, mean number of quakes in time period is: 4939.65, Number of days in period = 7304.84

magnitude cutoff = 7., years = 30., Probability= 3.79499*10^-7 per quake or 3.14958*10^-7 per day, count = 1415.49, sd = 308.254, mean number of quakes in time period is: 8184.41, Number of days in period = 10957.3

magnitude cutoff = 7.5, years = 0.0142664, Probability= 3.25774*10^-7 per quake or 3.66524*10^-7 per day, count = 0.9605, sd = 0.904119, mean number of quakes in time period is: 0.651389, Number of days in period = 5.21071

magnitude cutoff = 7.5, years = 10., Probability= 1.64764*10^-7 per quake or 1.10588*10^-7 per day, count = 182.158, sd = 44.0865, mean number of quakes in time period is: 2206.33, Number of days in period = 3652.42

magnitude cutoff = 7.5, years = 20., Probability= 1.66161*10^-7 per quake or 1.24845*10^-7 per day, count = 411.323, sd = 90.761, mean number of quakes in time period is: 4939.65, Number of days in period = 7304.84

magnitude cutoff = 7.5, years = 30., Probability= 1.66633*10^-7 per quake or 1.38294*10^-7 per day, count = 683.458, sd = 149.198, mean number of quakes in time period is: 8184.41, Number of days in period = 10957.3

magnitude cutoff = 8., years = 0.041706, Probability= 1.23749*10^-7 per quake or 1.19055*10^-7 per day, count = 0.945, sd = 0.900642, mean number of quakes in time period is: 1.62833, Number of days in period = 15.2328

magnitude cutoff = 8., years = 10., Probability= 5.23464*10^-8 per quake or 3.51345*10^-8 per day, count = 60.1813, sd = 14.8164, mean number of quakes in time period is: 2206.33, Number of days in period = 3652.42

magnitude cutoff = 8., years = 20., Probability= 5.24572*10^-8 per quake or 3.94138*10^-8 per day, count = 135.022, sd = 29.9889, mean number of quakes in time period is: 4939.65, Number of days in period = 7304.84

magnitude cutoff = 8., years = 30., Probability= 5.24986*10^-8 per quake or 4.35703*10^-8 per day, count = 223.893, sd = 49.0215, mean number of quakes in time period is: 8184.41, Number of days in period = 10957.3

magnitude cutoff = 8.5, years = 0.188188, Probability= 3.42762*10^-8 per quake or 2.67478*10^-8 per day, count = 0.958, sd = 0.890454, mean number of quakes in time period is: 5.95972, Number of days in period = 68.7342

magnitude cutoff = 8.5, years = 10., Probability= 1.7296*10^-8 per quake or 1.1609*10^-8 per day, count = 19.8848, sd = 4.94735, mean number of quakes in time period is: 2206.33, Number of days in period = 3652.42

magnitude cutoff = 8.5, years = 20., Probability= 1.72456*10^-8 per quake or 1.29575*10^-8 per day, count = 44.3893, sd = 9.9018, mean number of quakes in time period is: 4939.65, Number of days in period = 7304.84

magnitude cutoff = 8.5, years = 30., Probability= 1.72334*10^-8 per quake or 1.43026*10^-8 per day, count = 73.4961, sd = 16.1398, mean number of quakes in time period is: 8184.41, Number of days in period = 10957.3

magnitude cutoff = 9., years = 30., Probability= 7.2059*10^-12 per quake or 5.25429*10^-12 per day, count = 0.03, sd = 0.0321016, mean number of quakes in time period is: 887.74, Number of days in period = 10957.3

magnitude cutoff = 9., years = 10., Probability= 8.56765*10^-12 per quake or 5.75053*10^-12 per day, count = 0.00985, sd = 0.0319137, mean number of quakes in time period is: 2206.33, Number of days in period = 3652.42

magnitude cutoff = 9., years = 20., Probability= 7.07084*10^-12 per quake or 5.31268*10^-12 per day, count = 0.0182, sd = 0.0434707, mean number of quakes in time period is: 4939.65, Number of days in period = 7304.84

magnitude cutoff = 9., years = 30., Probability= 6.81168*10^-12 per quake or 5.65323*10^-12 per day, count = 0.02905, sd = 0.0559702, mean number of quakes in time period is: 8184.41, Number of days in period = 10957.3

California

magnitude cutoff = 5., years = 0.031119, Probability= 4.10051*10-6 per quake or 1.11027*10-6 per day, count = 0.965, sd = 0.886568, mean number of quakes in time period is: 0.341944, Number of days in period = 11.366

magnitude cutoff = 5., years = 10., Probability= 2.1294*10-6 per quake or 4.97301*10-7 per day, count = 119.507, sd = 18.6931, mean number of quakes in time period is: 767.687, Number of days in period = 3652.42

magnitude cutoff = 5., years = 20., Probability= 2.09242*10-6 per quake or 5.16679*10-7 per day, count = 248.369, sd = 33.5911, mean number of quakes in time period is: 1623.4, Number of days in period = 7304.84

magnitude cutoff = 5., years = 30., Probability= 2.0772*10-6 per quake or 5.42477*10-7 per day, count = 391.093, sd = 48.1261, mean number of quakes in time period is: 2575.42, Number of days in period = 10957.3

magnitude cutoff = 5.5, years = 0.0938086, Probability= 1.37694*10-6 per quake or 3.24113*10-7 per day, count = 0.956, sd = 0.878644, mean number of quakes in time period is: 0.896111, Number of days in period = 34.2629

magnitude cutoff = 5.5, years = 10., Probability= 6.09263*10-7 per quake or 1.42287*10-7 per day, count = 40.0142, sd = 6.46341, mean number of quakes in time period is: 767.687, Number of days in period = 3652.42

magnitude cutoff = 5.5, years = 20., Probability= 5.94572*10-7 per quake or 1.46817*10-7 per day, count = 82.7159, sd = 11.4786, mean number of quakes in time period is: 1623.4, Number of days in period = 7304.84

magnitude cutoff = 5.5, years = 30., Probability= 5.88655*10-7 per quake or 1.53732*10-7 per day, count = 129.965, sd = 16.3178, mean number of quakes in time period is: 2575.42, Number of days in period = 10957.3

magnitude cutoff = 6., years = 0.433951, Probability= 3.13288*10-7 per quake or 6.88286*10-8 per day, count = 0.9645, sd = 0.877045, mean number of quakes in time period is: 3.86906, Number of days in period = 158.497

magnitude cutoff = 6., years = 10., Probability= 1.7457*10-7 per quake or 4.07691*10-8 per day, count = 12.0283, sd = 2.01537, mean number of quakes in time period is: 767.687, Number of days in period = 3652.42

magnitude cutoff = 6., years = 20., Probability= 1.69619*10-7 per quake or 4.18839*10-8 per day, count = 24.7329, sd = 3.53038, mean number of quakes in time period is: 1623.4, Number of days in period = 7304.84

magnitude cutoff = 6., years = 30., Probability= 1.67678*10-7 per quake or 4.37906*10-8 per day, count = 38.7987, sd = 4.97618, mean number of quakes in time period is: 2575.42, Number of days in period = 10957.3

magnitude cutoff = 6.5, years = 1.36343, Probability= 9.60229*10-8 per quake or 2.04181*10-8 per day, count = 0.9395, sd = 0.85297, mean number of quakes in time period is: 11.7656, Number of days in period = 497.982

magnitude cutoff = 6.5, years = 10., Probability= 7.52768*10-8 per quake or 1.75801*10-8 per day, count = 5.33965, sd = 1.02292, mean number of quakes in time period is: 767.687, Number of days in period = 3652.42

magnitude cutoff = 6.5, years = 20., Probability= 7.32186*10-8 per quake or 1.80798*10-8 per day, count = 10.9829, sd = 1.70657, mean number of quakes in time period is: 1623.4, Number of days in period = 7304.84

magnitude cutoff = 6.5, years = 30., Probability= 7.2337*10-8 per quake or 1.88914*10-8 per day, count = 17.2138, sd = 2.37659, mean number of quakes in time period is: 2575.42, Number of days in period = 10957.3

magnitude cutoff = 7., years = 7.67332, Probability= 1.76445*10-8 per quake or 3.68784*10-9 per day, count = 0.955, sd = 0.86679, mean number of quakes in time period is: 65.0854, Number of days in period = 2802.62

magnitude cutoff = 7., years = 10., Probability= 1.75657*10-8 per quake or 4.1023*10-9 per day, count = 1.246, sd = 0.402275, mean number of quakes in time period is: 767.687, Number of days in period = 3652.42

magnitude cutoff = 7., years = 20., Probability= 1.66459*10-8 per quake or 4.11037*10-9 per day, count = 2.4969, sd = 0.607466, mean number of quakes in time period is: 1623.4, Number of days in period = 7304.84

magnitude cutoff = 7., years = 30., Probability= 1.63439*10-8 per quake or 4.26835*10-9 per day, count = 3.8893, sd = 0.78934, mean number of quakes in time period is: 2575.42, Number of days in period = 10957.3

magnitude cutoff = 7.5, years = 30., Probability= 0. per quake or 0. per day, count = 0., sd = 0., mean number of quakes in time period is: 288.514, Number of days in period = 10957.3

Vancouver-Seattle

magnitude cutoff = 5., years = 0.265091, Probability= 1.05536*10-6 per quake or 1.14531*10-7 per day, count = 1.0045, sd = 0.921431, mean number of quakes in time period is: 1.1675, Number of days in period = 96.8223

magnitude cutoff = 5., years = 10., Probability= 6.81604*10-7 per quake or 6.21468*10-8 per day, count = 18.341, sd = 3.24208, mean number of quakes in time period is: 299.716, Number of days in period = 3652.42

magnitude cutoff = 5., years = 20., Probability= 6.66746*10-7 per quake or 5.99139*10-8 per day, count = 35.3373, sd = 4.92424, mean number of quakes in time period is: 590.772, Number of days in period = 7304.84

magnitude cutoff = 5., years = 30., Probability= 6.59009*10-7 per quake or 5.93262*10-8 per day, count = 52.4761, sd = 6.42654, mean number of quakes in time period is: 887.769, Number of days in period = 10957.3

magnitude cutoff = 5.5, years = 0.462771, Probability= 6.08549*10-7 per quake or 6.42578*10-8 per day, count = 1.0305, sd = 0.9404, mean number of quakes in time period is: 1.98306, Number of days in period = 169.023

magnitude cutoff = 5.5, years = 10., Probability= 4.07805*10-7 per quake or 3.71825*10-8 per day, count = 11.6462, sd = 2.06304, mean number of quakes in time period is: 299.716, Number of days in period = 3652.42

magnitude cutoff = 5.5, years = 20., Probability= 3.97827*10-7 per quake or 3.57487*10-8 per day, count = 22.4222, sd = 3.12887, mean number of quakes in time period is: 590.772, Number of days in period = 7304.84

magnitude cutoff = 5.5, years = 30., Probability= 3.92949*10-7 per quake or 3.53746*10-8 per day, count = 33.2913, sd = 4.07877, mean number of quakes in time period is: 887.769, Number of days in period = 10957.3

magnitude cutoff = 6., years = 1.01586, Probability= 2.91341*10-7 per quake or 2.90768*10-8 per day, count = 1.037, sd = 0.941742, mean number of quakes in time period is: 4.1145, Number of days in period = 371.035

magnitude cutoff = 6., years = 10., Probability= 2.13727*10-7 per quake or 1.94871*10-8 per day, count = 6.17745, sd = 1.11133, mean number of quakes in time period is: 299.716, Number of days in period = 3652.42

magnitude cutoff = 6., years = 20., Probability= 2.08272*10-7 per quake or 1.87154*10-8 per day, count = 11.8856, sd = 1.68396, mean number of quakes in time period is: 590.772, Number of days in period = 7304.84

magnitude cutoff = 6., years = 30., Probability= 2.05554*10-7 per quake or 1.85047*10-8 per day, count = 17.6403, sd = 2.18363, mean number of quakes in time period is: 887.769, Number of days in period = 10957.3

magnitude cutoff = 6.5, years = 4.07019, Probability= 7.12557*10-8 per quake or 6.17058*10-9 per day, count = 0.932, sd = 0.845888, mean number of quakes in time period is: 14.3041, Number of days in period = 1486.6

magnitude cutoff = 6.5, years = 10., Probability= 6.84619*10-8 per quake or 6.24217*10-9 per day, count = 2.08475, sd = 0.529767, mean number of quakes in time period is: 299.716, Number of days in period = 3652.42

magnitude cutoff = 6.5, years = 20., Probability= 6.64633*10-8 per quake or 5.97239*10-9 per day, count = 3.9893, sd = 0.784036, mean number of quakes in time period is: 590.772, Number of days in period = 7304.84

magnitude cutoff = 6.5, years = 30., Probability= 6.56104*10-8 per quake or 5.90647*10-9 per day, count = 5.9179, sd = 0.972107, mean number of quakes in time period is: 887.769, Number of days in period = 10957.3

magnitude cutoff = 7., years = 27.6656, Probability= 1.16838*10-8 per quake or 9.38992*10-10 per day, count = 0.964, sd = 0.87371, mean number of quakes in time period is: 90.2312, Number of days in period = 10104.6

magnitude cutoff = 7., years = 10., Probability= 1.21259*10-8 per quake or 1.10561*10-9 per day, count = 0.36925, sd = 0.195521, mean number of quakes in time period is: 299.716, Number of days in period = 3652.42

magnitude cutoff = 7., years = 20., Probability= 1.19405*10-8 per quake or 1.07297*10-9 per day, count = 0.7167, sd = 0.280696, mean number of quakes in time period is: 590.772, Number of days in period = 7304.84

magnitude cutoff = 7., years = 30., Probability= 1.18495*10-8 per quake or 1.06674*10-9 per day, count = 1.0688, sd = 0.347602, mean number of quakes in time period is: 887.769, Number of days in period = 10957.3

magnitude cutoff = 7.5, years = 30., Probability= 0. per quake or 0. per day, count = 0., sd = 0., mean number of quakes in time period is: 98.0637, Number of days in period = 10957.3

Country = Saudi Arabia

magnitude cutoff = 4.5, years = 0.0196344, Probability= 3.42405*10^-6 per quake or 9.94797*10^-7 per day, count = 1.039, sd = 0.952287, mean number of quakes in time period is: 0.2315, Number of days in period = 7.17132

magnitude cutoff = 4.5, years = 10., Probability= 2.92091*10^-6 per quake or 3.9789*10^-7 per day, count = 175.519, sd = 29.8057, mean number of quakes in time period is: 447.784, Number of days in period = 3652.42

magnitude cutoff = 4.5, years = 20., Probability= 2.94756*10^-6 per quake or 3.80846*10^-7 per day, count = 335.629, sd = 45.222, mean number of quakes in time period is: 849.455, Number of days in period = 7304.84

magnitude cutoff = 4.5, years = 30., Probability= 2.95936*10^-6 per quake or 3.75849*10^-7 per day, count = 496.61, sd = 56.22, mean number of quakes in time period is: 1252.45, Number of days in period = 10957.3

magnitude cutoff = 5., years = 0.0331616, Probability= 2.2365*10^-6 per quake or 5.51738*10^-7 per day, count = 1.117, sd = 1.02786, mean number of quakes in time period is: 0.332, Number of days in period = 12.112

magnitude cutoff = 5., years = 10., Probability= 1.53182*10^-6 per quake or 2.08666*10^-7 per day, count = 111.664, sd = 20.4834, mean number of quakes in time period is: 447.784, Number of days in period = 3652.42

magnitude cutoff = 5., years = 20., Probability= 1.53997*10^-6 per quake or 1.98975*10^-7 per day, count = 212.87, sd = 30.6915, mean number of quakes in time period is: 849.455, Number of days in period = 7304.84

magnitude cutoff = 5., years = 30., Probability= 1.54377*10^-6 per quake or 1.96064*10^-7 per day, count = 314.587, sd = 37.9633, mean number of quakes in time period is: 1252.45, Number of days in period = 10957.3

magnitude cutoff = 5.5, years = 0.0657181, Probability= 1.27178*10^-6 per quake or 2.68577*10^-7 per day, count = 1.1445, sd = 1.0586, mean number of quakes in time period is: 0.563222, Number of days in period = 24.003

magnitude cutoff = 5.5, years = 10., Probability= 7.51098*10^-7 per quake or 1.02316*10^-7 per day, count = 59.5009, sd = 11.535, mean number of quakes in time period is: 447.784, Number of days in period = 3652.42

magnitude cutoff = 5.5, years = 20., Probability= 7.51134*10^-7 per quake or 9.7052*10^-8 per day, count = 112.906, sd = 17.1326, mean number of quakes in time period is: 849.455, Number of days in period = 7304.84

magnitude cutoff = 5.5, years = 30., Probability= 7.51678*10^-7 per quake or 9.54657*10^-8 per day, count = 166.602, sd = 21.145, mean number of quakes in time period is: 1252.45, Number of days in period = 10957.3

magnitude cutoff = 6., years = 0.199561, Probability= 4.7069*10^-7 per quake or 8.61748*10^-8 per day, count = 1.121, sd = 1.03434, mean number of quakes in time period is: 1.48272, Number of days in period = 72.8882

magnitude cutoff = 6., years = 10., Probability= 2.60247*10^-7 per quake or 3.54512*10^-8 per day, count = 20.6743, sd = 4.15145, mean number of quakes in time period is: 447.784, Number of days in period = 3652.42

magnitude cutoff = 6., years = 20., Probability= 2.59438*10^-7 per quake or 3.35212*10^-8 per day, count = 39.0815, sd = 6.09277, mean number of quakes in time period is: 849.455, Number of days in period = 7304.84

magnitude cutoff = 6., years = 30., Probability= 2.59047*10^-7 per quake or 3.28998*10^-8 per day, count = 57.5246, sd = 7.51658, mean number of quakes in time period is: 1252.45, Number of days in period = 10957.3

magnitude cutoff = 6.5, years = 3.97281, Probability= 2.91566*10^-8 per quake or 3.91653*10^-9 per day, count = 1.06, sd = 0.961734, mean number of quakes in time period is: 21.6572, Number of days in period = 1451.04

magnitude cutoff = 6.5, years = 10., Probability= 2.66892*10^-8 per quake or 3.63564*10^-9 per day, count = 2.2291, sd = 0.552717, mean number of quakes in time period is: 447.784, Number of days in period = 3652.42

magnitude cutoff = 6.5, years = 20., Probability= 2.6633*10^-8 per quake or 3.44118*10^-9 per day, count = 4.21975, sd = 0.777636, mean number of quakes in time period is: 849.455, Number of days in period = 7304.84

magnitude cutoff = 6.5, years = 30., Probability= 2.65543*10^-8 per quake or 3.37248*10^-9 per day, count = 6.20325, sd = 0.960196, mean number of quakes in time period is: 1252.45, Number of days in period = 10957.3

magnitude cutoff = 7., years = 30., Probability= 0. per quake or 0. per day, count = 0., sd = 0., mean number of quakes in time period is: 139.074, Number of days in period = 10957.3

Germany-Austria-Czech Republic

magnitude cutoff = 5., years = 0.0535671, Probability= 4.80181*10-7 per quake or 2.0001*10-7 per day, count = 1.1086, sd = 1.0255, mean number of quakes in time period is: 1.1642, Number of days in period = 19.565

magnitude cutoff = 5., years = 10., Probability= 2.68991*10-7 per quake or 7.29742*10-8 per day, count = 65.2631, sd = 12.7502, mean number of quakes in time period is: 867.002, Number of days in period = 3652.42

magnitude cutoff = 5., years = 20., Probability= 2.71275*10-7 per quake or 7.04155*10-8 per day, count = 125.895, sd = 18.5451, mean number of quakes in time period is: 1659.12, Number of days in period = 7304.84

magnitude cutoff = 5., years = 30., Probability= 2.71609*10-7 per quake or 6.95731*10-8 per day, count = 186.557, sd = 22.4841, mean number of quakes in time period is: 2455.88, Number of days in period = 10957.3

magnitude cutoff = 5.5, years = 2.51598, Probability= 1.34752*10-8 per quake or 3.65766*10-9 per day, count = 1.0026, sd = 0.88918, mean number of quakes in time period is: 35.6334, Number of days in period = 918.944

magnitude cutoff = 5.5, years = 10., Probability= 1.27112*10-8 per quake or 3.4484*10-9 per day, count = 3.28733, sd = 0.76614, mean number of quakes in time period is: 867.002, Number of days in period = 3652.42

magnitude cutoff = 5.5, years = 20., Probability= 1.27081*10-8 per quake or 3.29866*10-9 per day, count = 6.28915, sd = 1.09861, mean number of quakes in time period is: 1659.12, Number of days in period = 7304.84

magnitude cutoff = 5.5, years = 30., Probability= 1.26904*10-8 per quake or 3.25067*10-9 per day, count = 9.2965, sd = 1.34767, mean number of quakes in time period is: 2455.88, Number of days in period = 10957.3

magnitude cutoff = 6., years = 30., Probability= 0. per quake or 0. per day, count = 0., sd = 0., mean number of quakes in time period is: 351.595, Number of days in period = 10957.3

magnitude cutoff = 6., years = 10., Probability= 0. per quake or 0. per day, count = 0., sd = 0., mean number of quakes in time period is: 867.002, Number of days in period = 3652.42

magnitude cutoff = 6., years = 20., Probability= 0. per quake or 0. per day, count = 0., sd = 0., mean number of quakes in time period is: 1659.12, Number of days in period = 7304.84

magnitude cutoff = 6., years = 30., Probability= 0. per quake or 0. per day, count = 0., sd = 0., mean number of quakes in time period is: 2455.88, Number of days in period = 10957.3

Italy-Switzerland

magnitude cutoff = 5., years = 0.0117222, Probability= 1.72473*10-6 per quake or 9.87513*10-7 per day, count = 1.0036, sd = 0.842837, mean number of quakes in time period is: 0.817133, Number of days in period = 4.28146

magnitude cutoff = 5., years = 10., Probability= 1.30543*10-6 per quake or 5.59967*10-7 per day, count = 347.124, sd = 118.579, mean number of quakes in time period is: 1175.04, Number of days in period = 3652.42

magnitude cutoff = 5., years = 20., Probability= 1.32457*10-6 per quake or 5.88316*10-7 per day, count = 728.885, sd = 256.819, mean number of quakes in time period is: 2433.38, Number of days in period = 7304.84

magnitude cutoff = 5., years = 30., Probability= 1.33309*10-6 per quake or 6.36979*10-7 per day, count = 1183.49, sd = 399.508, mean number of quakes in time period is: 3926.72, Number of days in period = 10957.3

magnitude cutoff = 5.5, years = 0.0183881, Probability= 1.00254*10-6 per quake or 5.67242*10-7 per day, count = 1.0156, sd = 0.899727, mean number of quakes in time period is: 1.26667, Number of days in period = 6.71611

magnitude cutoff = 5.5, years = 10., Probability= 6.21799*10-7 per quake or 2.66723*10-7 per day, count = 192.974, sd = 67.6433, mean number of quakes in time period is: 1175.04, Number of days in period = 3652.42

magnitude cutoff = 5.5, years = 20., Probability= 6.28535*10-7 per quake or 2.79169*10-7 per day, count = 403.881, sd = 143.863, mean number of quakes in time period is: 2433.38, Number of days in period = 7304.84

magnitude cutoff = 5.5, years = 30., Probability= 6.31833*10-7 per quake or 3.01904*10-7 per day, count = 655.105, sd = 222.674, mean number of quakes in time period is: 3926.72, Number of days in period = 10957.3

magnitude cutoff = 6., years = 0.0412681, Probability= 4.39985*10-7 per quake or 2.33945*10-7 per day, count = 1.0096, sd = 0.99093, mean number of quakes in time period is: 2.67147, Number of days in period = 15.0729

magnitude cutoff = 6., years = 10., Probability= 2.30588*10-7 per quake or 9.89114*10-8 per day, count = 77.676, sd = 27.5834, mean number of quakes in time period is: 1175.04, Number of days in period = 3652.42

magnitude cutoff = 6., years = 20., Probability= 2.31875*10-7 per quake or 1.02989*10-7 per day, count = 161.759, sd = 57.9454, mean number of quakes in time period is: 2433.38, Number of days in period = 7304.84

magnitude cutoff = 6., years = 30., Probability= 2.32599*10-7 per quake or 1.11141*10-7 per day, count = 261.845, sd = 89.3343, mean number of quakes in time period is: 3926.72, Number of days in period = 10957.3

magnitude cutoff = 6.5, years = 0.507511, Probability= 4.56507*10-8 per quake or 1.88104*10-8 per day, count = 1.0164, sd = 0.856391, mean number of quakes in time period is: 25.4599, Number of days in period = 185.364

magnitude cutoff = 6.5, years = 10., Probability= 3.17635*10-8 per quake or 1.36251*10-8 per day, count = 10.8798, sd = 3.91237, mean number of quakes in time period is: 1175.04, Number of days in period = 3652.42

magnitude cutoff = 6.5, years = 20., Probability= 3.17504*10-8 per quake or 1.41022*10-8 per day, count = 22.5216, sd = 8.09739, mean number of quakes in time period is: 2433.38, Number of days in period = 7304.84

magnitude cutoff = 6.5, years = 30., Probability= 3.17893*10-8 per quake or 1.51896*10-8 per day, count = 36.3874, sd = 12.4429, mean number of quakes in time period is: 3926.72, Number of days in period = 10957.3

magnitude cutoff = 7., years = 30., Probability= 0. per quake or 0. per day, count = 0., sd = 0., mean number of quakes in time period is: 1315.93, Number of days in period = 10957.3

magnitude cutoff = 7., years = 10., Probability= 0. per quake or 0. per day, count = 0., sd = 0., mean number of quakes in time period is: 1175.04, Number of days in period = 3652.42

magnitude cutoff = 7., years = 20., Probability= 0. per quake or 0. per day, count = 0., sd = 0., mean number of quakes in time period is: 2433.38, Number of days in period = 7304.84

magnitude cutoff = 7., years = 30., Probability= 0. per quake or 0. per day, count = 0., sd = 0., mean number of quakes in time period is: 3926.72, Number of days in period = 10957.3

magnitude cutoff = 7.5, years = 30., Probability= 0. per quake or 0. per day, count = 0., sd = 0., mean number of quakes in time period is: 1315.93, Number of days in period = 10957.3

Mexico

magnitude cutoff = 5., years = 0.0060782, Probability= 1.7442*10-6 per quake or 1.51211*10-6 per day, count = 1.0302, sd = 0.856922, mean number of quakes in time period is: 0.641533, Number of days in period = 2.22001

magnitude cutoff = 5., years = 10., Probability= 1.44319*10-6 per quake or 6.60575*10-7 per day, count = 519.05, sd = 163.281, mean number of quakes in time period is: 1253.83, Number of days in period = 3652.42

magnitude cutoff = 5., years = 20., Probability= 1.45356*10-6 per quake or 6.55754*10-7 per day, count = 1028.07, sd = 225.086, mean number of quakes in time period is: 2471.61, Number of days in period = 7304.84

magnitude cutoff = 5., years = 30., Probability= 1.44113*10-6 per quake or 6.47719*10-7 per day, count = 1521.93, sd = 287.442, mean number of quakes in time period is: 3693.57, Number of days in period = 10957.3

magnitude cutoff = 5.5, years = 0.00824855, Probability= 1.2463*10-6 per quake or 1.05116*10-6 per day, count = 1.0604, sd = 0.903376, mean number of quakes in time period is: 0.847, Number of days in period = 3.01272

magnitude cutoff = 5.5, years = 10., Probability= 8.55162*10-7 per quake or 3.91423*10-7 per day, count = 341.551, sd = 117.962, mean number of quakes in time period is: 1253.83, Number of days in period = 3652.42

magnitude cutoff = 5.5, years = 20., Probability= 8.55461*10-7 per quake or 3.85929*10-7 per day, count = 672.222, sd = 160.519, mean number of quakes in time period is: 2471.61, Number of days in period = 7304.84

magnitude cutoff = 5.5, years = 30., Probability= 8.45805*10-7 per quake or 3.80148*10-7 per day, count = 992.598, sd = 202.782, mean number of quakes in time period is: 3693.57, Number of days in period = 10957.3

magnitude cutoff = 6., years = 0.0120464, Probability= 7.9613*10-7 per quake or 6.46189*10-7 per day, count = 1.035, sd = 0.92168, mean number of quakes in time period is: 1.1904, Number of days in period = 4.39986

magnitude cutoff = 6., years = 10., Probability= 4.63133*10-7 per quake or 2.11984*10-7 per day, count = 208.3, sd = 76.0192, mean number of quakes in time period is: 1253.83, Number of days in period = 3652.42

magnitude cutoff = 6., years = 20., Probability= 4.58509*10-7 per quake or 2.0685*10-7 per day, count = 406.31, sd = 102.645, mean number of quakes in time period is: 2471.61, Number of days in period = 7304.84

magnitude cutoff = 6., years = 30., Probability= 4.51742*10-7 per quake or 2.03036*10-7 per day, count = 598.145, sd = 128.835, mean number of quakes in time period is: 3693.57, Number of days in period = 10957.3

magnitude cutoff = 6.5, years = 0.0195717, Probability= 5.07479*10-7 per quake or 3.89278*10-7 per day, count = 1.0166, sd = 0.969414, mean number of quakes in time period is: 1.8278, Number of days in period = 7.14839

magnitude cutoff = 6.5, years = 10., Probability= 2.51166*10-7 per quake or 1.14963*10-7 per day, count = 113.234, sd = 42.7155, mean number of quakes in time period is: 1253.83, Number of days in period = 3652.42

magnitude cutoff = 6.5, years = 20., Probability= 2.46348*10-7 per quake or 1.11136*10-7 per day, count = 218.746, sd = 57.2696, mean number of quakes in time period is: 2471.61, Number of days in period = 7304.84

magnitude cutoff = 6.5, years = 30., Probability= 2.42086*10-7 per quake or 1.08806*10-7 per day, count = 321.161, sd = 71.4552, mean number of quakes in time period is: 3693.57, Number of days in period = 10957.3

magnitude cutoff = 7., years = 0.0348961, Probability= 3.00222*10-7 per quake or 2.1828*10-7 per day, count = 1.0294, sd = 1.05266, mean number of quakes in time period is: 3.08893, Number of days in period = 12.7455

magnitude cutoff = 7., years = 10., Probability= 1.29309*10-7 per quake or 5.91869*10-8 per day, count = 59.3048, sd = 22.7366, mean number of quakes in time period is: 1253.83, Number of days in period = 3652.42

magnitude cutoff = 7., years = 20., Probability= 1.26*10-7 per quake or 5.68433*10-8 per day, count = 113.8, sd = 30.485, mean number of quakes in time period is: 2471.61, Number of days in period = 7304.84

magnitude cutoff = 7., years = 30., Probability= 1.23574*10-7 per quake or 5.55407*10-8 per day, count = 166.735, sd = 37.8672, mean number of quakes in time period is: 3693.57, Number of days in period = 10957.3

magnitude cutoff = 7.5, years = 0.130559, Probability= 9.87637*10-8 per quake or 6.02525*10-8 per day, count = 1.1212, sd = 1.10939, mean number of quakes in time period is: 9.69713, Number of days in period = 47.6856

magnitude cutoff = 7.5, years = 10., Probability= 4.23031*10-8 per quake or 1.93629*10-8 per day, count = 20.6904, sd = 8.10272, mean number of quakes in time period is: 1253.83, Number of days in period = 3652.42

magnitude cutoff = 7.5, years = 20., Probability= 4.07312*10-8 per quake or 1.83753*10-8 per day, count = 39.2638, sd = 10.8333, mean number of quakes in time period is: 2471.61, Number of days in period = 7304.84

magnitude cutoff = 7.5, years = 30., Probability= 3.98033*10-8 per quake or 1.78897*10-8 per day, count = 57.336, sd = 13.3958, mean number of quakes in time period is: 3693.57, Number of days in period = 10957.3

magnitude cutoff = 8., years = 6.99321, Probability= 2.74019*10-9 per quake or 9.47766*10-10 per day, count = 0.9428, sd = 0.760083, mean number of quakes in time period is: 294.48, Number of days in period = 2554.21

magnitude cutoff = 8., years = 10., Probability= 2.51323*10-9 per quake or 1.15035*10-9 per day, count = 1.22725, sd = 0.666565, mean number of quakes in time period is: 1253.83, Number of days in period = 3652.42

magnitude cutoff = 8., years = 20., Probability= 2.20265*10-9 per quake or 9.93696*10-10 per day, count = 2.12025, sd = 0.880572, mean number of quakes in time period is: 2471.61, Number of days in period = 7304.84

magnitude cutoff = 8., years = 30., Probability= 2.08597*10-9 per quake or 9.37542*10-10 per day, count = 3.00065, sd = 1.06559, mean number of quakes in time period is: 3693.57, Number of days in period = 10957.3

Pakistan-Afghanistan

magnitude cutoff = 5., years = 0.00809302, Probability= 3.05978*10-6 per quake or 1.72661*10-6 per day, count = 1.0425, sd = 0.948346, mean number of quakes in time period is: 0.185333, Number of days in period = 2.95591

magnitude cutoff = 5., years = 10., Probability= 2.93722*10-6 per quake or 7.16101*10-7 per day, count = 434.683, sd = 97.4256, mean number of quakes in time period is: 801.422, Number of days in period = 3652.42

magnitude cutoff = 5., years = 20., Probability= 2.99143*10-6 per quake or 7.51365*10-7 per day, count = 911.315, sd = 188.333, mean number of quakes in time period is: 1651.3, Number of days in period = 7304.84

magnitude cutoff = 5., years = 30., Probability= 2.99032*10-6 per quake or 7.95315*10-7 per day, count = 1446.93, sd = 300.455, mean number of quakes in time period is: 2622.81, Number of days in period = 10957.3

magnitude cutoff = 5.5, years = 0.0104339, Probability= 2.19537*10-6 per quake or 1.2184*10-6 per day, count = 1.0915, sd = 0.994463, mean number of quakes in time period is: 0.235, Number of days in period = 3.81091

magnitude cutoff = 5.5, years = 10., Probability= 1.83951*10-6 per quake or 4.48476*10-7 per day, count = 321.402, sd = 76.0958, mean number of quakes in time period is: 801.422, Number of days in period = 3652.42

magnitude cutoff = 5.5, years = 20., Probability= 1.87272*10-6 per quake or 4.70376*10-7 per day, count = 673.416, sd = 143.15, mean number of quakes in time period is: 1651.3, Number of days in period = 7304.84

magnitude cutoff = 5.5, years = 30., Probability= 1.87275*10-6 per quake or 4.98083*10-7 per day, count = 1069.42, sd = 225.615, mean number of quakes in time period is: 2622.81, Number of days in period = 10957.3

magnitude cutoff = 6., years = 0.0137542, Probability= 1.6025*10-6 per quake or 8.71807*10-7 per day, count = 1.1115, sd = 1.01617, mean number of quakes in time period is: 0.303667, Number of days in period = 5.02363

magnitude cutoff = 6., years = 10., Probability= 1.10131*10-6 per quake or 2.68502*10-7 per day, count = 219.49, sd = 54.4555, mean number of quakes in time period is: 801.422, Number of days in period = 3652.42

magnitude cutoff = 6., years = 20., Probability= 1.11721*10-6 per quake or 2.80614*10-7 per day, count = 458.503, sd = 99.8862, mean number of quakes in time period is: 1651.3, Number of days in period = 7304.84

magnitude cutoff = 6., years = 30., Probability= 1.11662*10-6 per quake or 2.96981*10-7 per day, count = 727.744, sd = 155.771, mean number of quakes in time period is: 2622.81, Number of days in period = 10957.3

magnitude cutoff = 6.5, years = 0.0206748, Probability= 1.07882*10-6 per quake or 5.69461*10-7 per day, count = 1.114, sd = 1.02509, mean number of quakes in time period is: 0.442889, Number of days in period = 7.5513

magnitude cutoff = 6.5, years = 10., Probability= 6.62207*10-7 per quake or 1.61448*10-7 per day, count = 133.577, sd = 34.3653, mean number of quakes in time period is: 801.422, Number of days in period = 3652.42

magnitude cutoff = 6.5, years = 20., Probability= 6.68688*10-7 per quake or 1.67956*10-7 per day, count = 277.67, sd = 61.8617, mean number of quakes in time period is: 1651.3, Number of days in period = 7304.84

magnitude cutoff = 6.5, years = 30., Probability= 6.67612*10-7 per quake or 1.7756*10-7 per day, count = 440.2, sd = 95.2436, mean number of quakes in time period is: 2622.81, Number of days in period = 10957.3

magnitude cutoff = 7., years = 0.037804, Probability= 5.69206*10-7 per quake or 2.83931*10-7 per day, count = 1.097, sd = 1.02119, mean number of quakes in time period is: 0.765278, Number of days in period = 13.8076

magnitude cutoff = 7., years = 10., Probability= 2.86234*10-7 per quake or 6.97844*10-8 per day, count = 63.5416, sd = 16.8504, mean number of quakes in time period is: 801.422, Number of days in period = 3652.42

magnitude cutoff = 7., years = 20., Probability= 2.87124*10-7 per quake or 7.21177*10-8 per day, count = 131.262, sd = 29.8116, mean number of quakes in time period is: 1651.3, Number of days in period = 7304.84

magnitude cutoff = 7., years = 30., Probability= 2.85984*10-7 per quake or 7.60613*10-8 per day, count = 207.636, sd = 45.3169, mean number of quakes in time period is: 2622.81, Number of days in period = 10957.3

magnitude cutoff = 7.5, years = 0.180396, Probability= 1.35484*10-7 per quake or 5.9244*10-8 per day, count = 1.171, sd = 1.08388, mean number of quakes in time period is: 3.20128, Number of days in period = 65.8884

magnitude cutoff = 7.5, years = 10., Probability= 6.6352*10-8 per quake or 1.61768*10-8 per day, count = 15.9522, sd = 4.3148, mean number of quakes in time period is: 801.422, Number of days in period = 3652.42

magnitude cutoff = 7.5, years = 20., Probability= 6.59899*10-8 per quake or 1.65748*10-8 per day, count = 32.6894, sd = 7.50376, mean number of quakes in time period is: 1651.3, Number of days in period = 7304.84

magnitude cutoff = 7.5, years = 30., Probability= 6.55493*10-8 per quake or 1.74337*10-8 per day, count = 51.575, sd = 11.3307, mean number of quakes in time period is: 2622.81, Number of days in period = 10957.3

magnitude cutoff = 8., years = 19.2099, Probability= 2.0109*10-9 per quake or 4.52537*10-10 per day, count = 0.9525, sd = 0.86438, mean number of quakes in time period is: 175.44, Number of days in period = 7016.28

magnitude cutoff = 8., years = 10., Probability= 1.83681*10-9 per quake or 4.47818*10-10 per day, count = 0.4416, sd = 0.226175, mean number of quakes in time period is: 801.422, Number of days in period = 3652.42

magnitude cutoff = 8., years = 20., Probability= 2.03706*10-9 per quake or 5.11654*10-10 per day, count = 1.0091, sd = 0.352078, mean number of quakes in time period is: 1651.3, Number of days in period = 7304.84

magnitude cutoff = 8., years = 30., Probability= 2.09148*10-9 per quake or 5.56257*10-10 per day, count = 1.6456, sd = 0.498814, mean number of quakes in time period is: 2622.81, Number of days in period = 10957.3

Korea

magnitude cutoff = 5., years = 1.57466, Probability= 0.0000250211 per quake or 9.48628*10-8 per day, count = 1.0655, sd = 0.970035, mean number of quakes in time period is: 0.242278, Number of days in period = 575.131

magnitude cutoff = 5., years = 10., Probability= 0.0000264758 per quake or 7.82807*10-8 per day, count = 4.94375, sd = 1.06802, mean number of quakes in time period is: 9.71915, Number of days in period = 3652.42

magnitude cutoff = 5., years = 20., Probability= 0.0000270693 per quake or 7.60841*10-8 per day, count = 9.60145, sd = 1.49046, mean number of quakes in time period is: 18.4787, Number of days in period = 7304.84

magnitude cutoff = 5., years = 30., Probability= 0.0000273 per quake or 7.53877*10-8 per day, count = 14.2513, sd = 1.81274, mean number of quakes in time period is: 27.2322, Number of days in period = 10957.3

magnitude cutoff = 5.5, years = 3.3251, Probability= 0.0000112023 per quake or 3.64579*10-8 per day, count = 1.032, sd = 0.937165, mean number of quakes in time period is: 0.439167, Number of days in period = 1214.47

magnitude cutoff = 5.5, years = 10., Probability= 0.0000115784 per quake or 3.42336*10-8 per day, count = 2.5965, sd = 0.642747, mean number of quakes in time period is: 9.71915, Number of days in period = 3652.42

magnitude cutoff = 5.5, years = 20., Probability= 0.0000117745 per quake or 3.30948*10-8 per day, count = 5.0176, sd = 0.891537, mean number of quakes in time period is: 18.4787, Number of days in period = 7304.84

magnitude cutoff = 5.5, years = 30., Probability= 0.0000118424 per quake or 3.27021*10-8 per day, count = 7.43145, sd = 1.08073, mean number of quakes in time period is: 27.2322, Number of days in period = 10957.3

magnitude cutoff = 6., years = 5.65682, Probability= 6.63282*10-6 per quake or 1.94319*10-8 per day, count = 1.0225, sd = 0.927017, mean number of quakes in time period is: 0.672556, Number of days in period = 2066.11

magnitude cutoff = 6., years = 10., Probability= 6.77178*10-6 per quake or 2.0022*10-8 per day, count = 1.6762, sd = 0.448045, mean number of quakes in time period is: 9.71915, Number of days in period = 3652.42

magnitude cutoff = 6., years = 20., Probability= 6.87963*10-6 per quake or 1.93367*10-8 per day, count = 3.23765, sd = 0.628844, mean number of quakes in time period is: 18.4787, Number of days in period = 7304.84

magnitude cutoff = 6., years = 30., Probability= 6.90608*10-6 per quake or 1.90708*10-8 per day, count = 4.7897, sd = 0.760081, mean number of quakes in time period is: 27.2322, Number of days in period = 10957.3

magnitude cutoff = 6.5, years = 12.2719, Probability= 3.28454*10-6 per quake or 8.66383*10-9 per day, count = 0.989, sd = 0.896375, mean number of quakes in time period is: 1.31367, Number of days in period = 4482.21

magnitude cutoff = 6.5, years = 10., Probability= 3.40932*10-6 per quake or 1.00803*10-8 per day, count = 0.8439, sd = 0.298994, mean number of quakes in time period is: 9.71915, Number of days in period = 3652.42

magnitude cutoff = 6.5, years = 20., Probability= 3.45113*10-6 per quake or 9.70014*10-9 per day, count = 1.62415, sd = 0.42849, mean number of quakes in time period is: 18.4787, Number of days in period = 7304.84

magnitude cutoff = 6.5, years = 30., Probability= 3.4653*10-6 per quake or 9.56924*10-9 per day, count = 2.40335, sd = 0.518264, mean number of quakes in time period is: 27.2322, Number of days in period = 10957.3

magnitude cutoff = 7., years = 30., Probability= 0. per quake or 0. per day, count = 0., sd = 0., mean number of quakes in time period is: 3.01956, Number of days in period = 10957.3

Thailand-Indonesia-Malaysia

magnitude cutoff = 5., years = 0.00988941, Probability= 0.000061988 per quake or 0.0000171615 per day, count = 1., sd = 0.908627, mean number of quakes in time period is: 0.111111, Number of days in period = 3.61203

magnitude cutoff = 5., years = 10., Probability= 0.0000227295 per quake or 3.27374*10-6 per day, count = 473.454, sd = 33.6126, mean number of quakes in time period is: 473.454, Number of days in period = 3652.42

magnitude cutoff = 5., years = 20., Probability= 0.0000211762 per quake or 3.12901*10-6 per day, count = 971.434, sd = 79.2002, mean number of quakes in time period is: 971.434, Number of days in period = 7304.84

magnitude cutoff = 5., years = 30., Probability= 0.0000205899 per quake or 3.12816*10-6 per day, count = 1498.23, sd = 136.699, mean number of quakes in time period is: 1498.23, Number of days in period = 10957.3

magnitude cutoff = 5.5, years = 0.00988941, Probability= 0.000061988 per quake or 0.0000171615 per day, count = 1., sd = 0.908627, mean number of quakes in time period is: 0.111111, Number of days in period = 3.61203 magnitude cutoff = 5.5, years = 10., Probability= 0.0000227295 per quake or 3.27374*10-6 per day, count = 473.454, sd = 33.6126, mean number of quakes in time period is: 473.454, Number of days in period = 3652.42

magnitude cutoff = 5.5, years = 20., Probability= 0.0000211762 per quake or 3.12901*10-6 per day, count = 971.434, sd = 79.2002, mean number of quakes in time period is: 971.434, Number of days in period = 7304.84

magnitude cutoff = 5.5, years = 30., Probability= 0.0000205899 per quake or 3.12816*10-6 per day, count = 1498.23, sd = 136.699, mean number of quakes in time period is: 1498.23, Number of days in period = 10957.3

magnitude cutoff = 6., years = 0.0111839, Probability= 0.0000233949 per quake or 6.42026*10-6 per day, count = 1.021, sd = 0.928275, mean number of quakes in time period is: 0.124556, Number of days in period = 4.08484

magnitude cutoff = 6., years = 10., Probability= 0.0000212347 per quake or 3.05845*10-6 per day, count = 471.969, sd = 33.5812, mean number of quakes in time period is: 473.454, Number of days in period = 3652.42

magnitude cutoff = 6., years = 20., Probability= 0.0000198698 per quake or 2.93598*10-6 per day, count = 968.78, sd = 79.0241, mean number of quakes in time period is: 971.434, Number of days in period = 7304.84

magnitude cutoff = 6., years = 30., Probability= 0.0000193673 per quake or 2.94241*10-6 per day, count = 1494.42, sd = 136.366, mean number of quakes in time period is: 1498.23, Number of days in period = 10957.3

magnitude cutoff = 6.5, years = 0.0152022, Probability= 5.12*10-6 per quake or 1.3306*10-6 per day, count = 1.049, sd = 0.955336, mean number of quakes in time period is: 0.160333, Number of days in period = 5.55249

magnitude cutoff = 6.5, years = 10., Probability= 6.13951*10-6 per quake or 8.84277*10-7 per day, count = 354.918, sd = 40.4077, mean number of quakes in time period is: 473.454, Number of days in period = 3652.42

magnitude cutoff = 6.5, years = 20., Probability= 6.18816*10-6 per quake or 9.14368*10-7 per day, count = 748.039, sd = 73.593, mean number of quakes in time period is: 971.434, Number of days in period = 7304.84

magnitude cutoff = 6.5, years = 30., Probability= 6.22787*10-6 per quake or 9.46178*10-7 per day, count = 1164.77, sd = 116.287, mean number of quakes in time period is: 1498.23, Number of days in period = 10957.3

magnitude cutoff = 7., years = 0.0219463, Probability= 2.97151*10-6 per quake or 7.18806*10-7 per day, count = 1.0675, sd = 0.975195, mean number of quakes in time period is: 0.215444, Number of days in period = 8.01572

magnitude cutoff = 7., years = 10., Probability= 2.84954*10-6 per quake or 4.1042*10-7 per day, count = 227.529, sd = 40.5784, mean number of quakes in time period is: 473.454, Number of days in period = 3652.42

magnitude cutoff = 7., years = 20., Probability= 2.98631*10-6 per quake or 4.4126*10-7 per day, count = 492.748, sd = 61.6301, mean number of quakes in time period is: 971.434, Number of days in period = 7304.84

magnitude cutoff = 7., years = 30., Probability= 3.02745*10-6 per quake or 4.5995*10-7 per day, count = 770.565, sd = 89.0926, mean number of quakes in time period is: 1498.23, Number of days in period = 10957.3

magnitude cutoff = 7.5, years = 0.0357918, Probability= 1.52238*10-6 per quake or 3.33004*10-7 per day, count = 1.066, sd = 0.979906, mean number of quakes in time period is: 0.317722, Number of days in period = 13.0727

magnitude cutoff = 7.5, years = 10., Probability= 1.10791*10-6 per quake or 1.59572*10-7 per day, count = 125.054, sd = 27.9278, mean number of quakes in time period is: 473.454, Number of days in period = 3652.42

magnitude cutoff = 7.5, years = 20., Probability= 1.18624*10-6 per quake or 1.7528*10-7 per day, count = 274.532, sd = 40.0622, mean number of quakes in time period is: 971.434, Number of days in period = 7304.84

magnitude cutoff = 7.5, years = 30., Probability= 1.20316*10-6 per quake or 1.82792*10-7 per day, count = 429.348, sd = 55.5579, mean number of quakes in time period is: 1498.23, Number of days in period = 10957.3

magnitude cutoff = 8., years = 0.0791972, Probability= 6.9684*10-7 per quake or 1.27679*10-7 per day, count = 0.971, sd = 0.90639, mean number of quakes in time period is: 0.588889, Number of days in period = 28.9262

magnitude cutoff = 8., years = 10., Probability= 4.02992*10-7 per quake or 5.80432*10-8 per day, count = 49.4879, sd = 12.2085, mean number of quakes in time period is: 473.454, Number of days in period = 3652.42

magnitude cutoff = 8., years = 20., Probability= 4.27791*10-7 per quake or 6.32108*10-8 per day, count = 107.731, sd = 17.2996, mean number of quakes in time period is: 971.434, Number of days in period = 7304.84

magnitude cutoff = 8., years = 30., Probability= 4.32655*10-7 per quake or 6.57317*10-8 per day, count = 168.013, sd = 23.5721, mean number of quakes in time period is: 1498.23, Number of days in period = 10957.3

China

magnitude cutoff = 5., years = 0.00620698, Probability= 3.45374*10-6 per quake or 2.41162*10-6 per day, count = 1.058, sd = 0.962587, mean number of quakes in time period is: 0.175889, Number of days in period = 2.26705

magnitude cutoff = 5., years = 10., Probability= 3.85093*10-6 per quake or 1.13112*10-6 per day, count = 623.55, sd = 84.4238, mean number of quakes in time period is: 965.534, Number of days in period = 3652.42

magnitude cutoff = 5., years = 20., Probability= 3.93043*10-6 per quake or 1.14093*10-6 per day, count = 1261.36, sd = 116.221, mean number of quakes in time period is: 1908.41, Number of days in period = 7304.84

magnitude cutoff = 5., years = 30., Probability= 3.93954*10-6 per quake or 1.13958*10-6 per day, count = 1890.04, sd = 147.936, mean number of quakes in time period is: 2852.63, Number of days in period = 10957.3

magnitude cutoff = 5.5, years = 0.0075957, Probability= 2.32731*10-6 per quake or 1.59012*10-6 per day, count = 1.068, sd = 0.973685, mean number of quakes in time period is: 0.210611, Number of days in period = 2.77427

magnitude cutoff = 5.5, years = 10., Probability= 2.23056*10-6 per quake or 6.55175*10-7 per day, count = 477.334, sd = 76.8469, mean number of quakes in time period is: 965.534, Number of days in period = 3652.42

magnitude cutoff = 5.5, years = 20., Probability= 2.29255*10-6 per quake or 6.65486*10-7 per day, count = 968.914, sd = 104.814, mean number of quakes in time period is: 1908.41, Number of days in period = 7304.84

magnitude cutoff = 5.5, years = 30., Probability= 2.30044*10-6 per quake or 6.65442*10-7 per day, count = 1452.35, sd = 131.411, mean number of quakes in time period is: 2852.63, Number of days in period = 10957.3

magnitude cutoff = 6., years = 0.00973617, Probability= 1.50657*10-6 per quake or 1.00218*10-6 per day, count = 1.0415, sd = 0.952584, mean number of quakes in time period is: 0.262833, Number of days in period = 3.55606

magnitude cutoff = 6., years = 10., Probability= 1.28136*10-6 per quake or 3.7637*10-7 per day, count = 344.943, sd = 62.6778, mean number of quakes in time period is: 965.534, Number of days in period = 3652.42

magnitude cutoff = 6., years = 20., Probability= 1.31612*10-6 per quake or 3.82046*10-7 per day, count = 699.69, sd = 85.0168, mean number of quakes in time period is: 1908.41, Number of days in period = 7304.84

magnitude cutoff = 6., years = 30., Probability= 1.31929*10-6 per quake or 3.81628*10-7 per day, count = 1048., sd = 105.006, mean number of quakes in time period is: 2852.63, Number of days in period = 10957.3

magnitude cutoff = 6.5, years = 0.0134215, Probability= 9.79092*10-7 per quake or 6.24155*10-7 per day, count = 1.0045, sd = 0.923947, mean number of quakes in time period is: 0.347222, Number of days in period = 4.90208

magnitude cutoff = 6.5, years = 10., Probability= 7.28521*10-7 per quake or 2.13986*10-7 per day, count = 227.886, sd = 45.007, mean number of quakes in time period is: 965.534, Number of days in period = 3652.42

magnitude cutoff = 6.5, years = 20., Probability= 7.45556*10-7 per quake or 2.16421*10-7 per day, count = 460.939, sd = 61.0377, mean number of quakes in time period is: 1908.41, Number of days in period = 7304.84

magnitude cutoff = 6.5, years = 30., Probability= 7.46138*10-7 per quake or 2.15834*10-7 per day, count = 689.511, sd = 74.7408, mean number of quakes in time period is: 2852.63, Number of days in period = 10957.3

magnitude cutoff = 7., years = 0.0203874, Probability= 6.54298*10-7 per quake or 3.89389*10-7 per day, count = 0.964, sd = 0.89339, mean number of quakes in time period is: 0.492389, Number of days in period = 7.44633

magnitude cutoff = 7., years = 10., Probability= 4.08642*10-7 per quake or 1.20029*10-7 per day, count = 129.37, sd = 27.0838, mean number of quakes in time period is: 965.534, Number of days in period = 3652.42

magnitude cutoff = 7., years = 20., Probability= 4.16008*10-7 per quake or 1.20759*10-7 per day, count = 260.304, sd = 36.8695, mean number of quakes in time period is: 1908.41, Number of days in period = 7304.84

magnitude cutoff = 7., years = 30., Probability= 4.156*10-7 per quake or 1.2022*10-7 per day, count = 388.693, sd = 44.7503, mean number of quakes in time period is: 2852.63, Number of days in period = 10957.3

magnitude cutoff = 7.5, years = 0.0483782, Probability= 3.14554*10-7 per quake or 1.69322*10-7 per day, count = 1.04, sd = 0.980862, mean number of quakes in time period is: 1.05683, Number of days in period = 17.6698

magnitude cutoff = 7.5, years = 10., Probability= 1.57417*10-7 per quake or 4.62376*10-8 per day, count = 52.714, sd = 11.4476, mean number of quakes in time period is: 965.534, Number of days in period = 3652.42

magnitude cutoff = 7.5, years = 20., Probability= 1.59315*10-7 per quake or 4.62462*10-8 per day, count = 105.479, sd = 15.6602, mean number of quakes in time period is: 1908.41, Number of days in period = 7304.84

magnitude cutoff = 7.5, years = 30., Probability= 1.58806*10-7 per quake or 4.59375*10-8 per day, count = 157.172, sd = 18.9036, mean number of quakes in time period is: 2852.63, Number of days in period = 10957.3

magnitude cutoff = 8., years = 0.748945, Probability= 2.89271*10-8 per quake or 9.75597*10-9 per day, count = 0.958, sd = 0.877312, mean number of quakes in time period is: 10.2507, Number of days in period = 273.546

magnitude cutoff = 8., years = 10., Probability= 1.85768*10-8 per quake or 5.4565*10-9 per day, count = 6.43875, sd = 1.49193, mean number of quakes in time period is: 965.534, Number of days in period = 3652.42

magnitude cutoff = 8., years = 20., Probability= 1.84798*10-8 per quake or 5.36434*10-9 per day, count = 12.66, sd = 2.0575, mean number of quakes in time period is: 1908.41, Number of days in period = 7304.84

magnitude cutoff = 8., years = 30., Probability= 1.83095*10-8 per quake or 5.29635*10-9 per day, count = 18.7493, sd = 2.47308, mean number of quakes in time period is: 2852.63, Number of days in period = 10957.3

India

magnitude cutoff = 5., years = 0.0334694, Probability= 0.0000116271 per quake or 1.51183*10-6 per day, count = 0.937, sd = 0.857492, mean number of quakes in time period is: 0.176611, Number of days in period = 12.2244 magnitude cutoff = 5., years = 10., Probability= 0.0000132723 per quake or 7.19042*10-7 per day, count = 106.381, sd = 14.7584, mean number of quakes in time period is: 178.086, Number of days in period = 3652.42

magnitude cutoff = 5., years = 20., Probability= 0.0000134094 per quake or 6.93332*10-7 per day, count = 205.152, sd = 22.3544, mean number of quakes in time period is: 339.927, Number of days in period = 7304.84

magnitude cutoff = 5., years = 30., Probability= 0.000013472 per quake or 6.85776*10-7 per day, count = 304.376, sd = 27.9649, mean number of quakes in time period is: 501.989, Number of days in period = 10957.3

magnitude cutoff = 5.5, years = 0.050097, Probability= 6.10745*10-6 per quake or 7.25649*10-7 per day, count = 0.956, sd = 0.877339, mean number of quakes in time period is: 0.241556, Number of days in period = 18.2975

magnitude cutoff = 5.5, years = 10., Probability= 5.74447*10-6 per quake or 3.11212*10-7 per day, count = 71.8928, sd = 10.914, mean number of quakes in time period is: 178.086, Number of days in period = 3652.42

magnitude cutoff = 5.5, years = 20., Probability= 5.80454*10-6 per quake or 3.00123*10-7 per day, count = 138.614, sd = 16.2865, mean number of quakes in time period is: 339.927, Number of days in period = 7304.84

magnitude cutoff = 5.5, years = 30., Probability= 5.83237*10-6 per quake or 2.96889*10-7 per day, count = 205.618, sd = 20.2573, mean number of quakes in time period is: 501.989, Number of days in period = 10957.3

magnitude cutoff = 6., years = 0.0988005, Probability= 3.45283*10-6 per quake or 3.39388*10-7 per day, count = 0.979, sd = 0.902868, mean number of quakes in time period is: 0.394111, Number of days in period = 36.0861

magnitude cutoff = 6., years = 10., Probability= 3.03795*10-6 per quake or 1.64584*10-7 per day, count = 42.5613, sd = 6.87215, mean number of quakes in time period is: 178.086, Number of days in period = 3652.42

magnitude cutoff = 6., years = 20., Probability= 3.06451*10-6 per quake or 1.5845*10-7 per day, count = 81.9246, sd = 10.1328, mean number of quakes in time period is: 339.927, Number of days in period = 7304.84

magnitude cutoff = 6., years = 30., Probability= 3.0757*10-6 per quake or 1.56565*10-7 per day, count = 121.397, sd = 12.5655, mean number of quakes in time period is: 501.989, Number of days in period = 10957.3

magnitude cutoff = 6.5, years = 0.202309, Probability= 1.72671*10-6 per quake or 1.39648*10-7 per day, count = 0.976, sd = 0.898345, mean number of quakes in time period is: 0.664, Number of days in period = 73.8917

magnitude cutoff = 6.5, years = 10., Probability= 1.43334*10-6 per quake or 7.76528*10-8 per day, count = 24.1435, sd = 4.0328, mean number of quakes in time period is: 178.086, Number of days in period = 3652.42

magnitude cutoff = 6.5, years = 20., Probability= 1.44322*10-6 per quake or 7.46216*10-8 per day, count = 46.4021, sd = 5.91564, mean number of quakes in time period is: 339.927, Number of days in period = 7304.84

magnitude cutoff = 6.5, years = 30., Probability= 1.44688*10-6 per quake or 7.36518*10-8 per day, count = 68.6985, sd = 7.31199, mean number of quakes in time period is: 501.989, Number of days in period = 10957.3

magnitude cutoff = 7., years = 0.82678, Probability= 5.41305*10-7 per quake or 3.48364*10-8 per day, count = 0.995, sd = 0.90635, mean number of quakes in time period is: 2.15933, Number of days in period = 301.975

magnitude cutoff = 7., years = 10., Probability= 4.81374*10-7 per quake or 2.60789*10-8 per day, count = 8.10835, sd = 1.42964, mean number of quakes in time period is: 178.086, Number of days in period = 3652.42

magnitude cutoff = 7., years = 20., Probability= 4.83355*10-7 per quake or 2.49918*10-8 per day, count = 15.5407, sd = 2.08883, mean number of quakes in time period is: 339.927, Number of days in period = 7304.84

magnitude cutoff = 7., years = 30., Probability= 4.8414*10-7 per quake or 2.46445*10-8 per day, count = 22.9871, sd = 2.57412, mean number of quakes in time period is: 501.989, Number of days in period = 10957.3

magnitude cutoff = 7.5, years = 1.6794, Probability= 2.83342*10-7 per quake or 1.67538*10-8 per day, count = 0.972, sd = 0.88216, mean number of quakes in time period is: 4.02989, Number of days in period = 613.388

magnitude cutoff = 7.5, years = 10., Probability= 2.74332*10-7 per quake or 1.48622*10-8 per day, count = 4.6209, sd = 0.853831, mean number of quakes in time period is: 178.086, Number of days in period = 3652.42

magnitude cutoff = 7.5, years = 20., Probability= 2.74778*10-7 per quake or 1.42074*10-8 per day, count = 8.8346, sd = 1.23881, mean number of quakes in time period is: 339.927, Number of days in period = 7304.84

magnitude cutoff = 7.5, years = 30., Probability= 2.75085*10-7 per quake or 1.40028*10-8 per day, count = 13.0611, sd = 1.5246, mean number of quakes in time period is: 501.989, Number of days in period = 10957.3

magnitude cutoff = 8., years = 30., Probability= 6.34778*10-11 per quake or 2.89468*10-12 per day, count = 0.003, sd = 0.00610071, mean number of quakes in time period is: 55.5186, Number of days in period = 10957.3

magnitude cutoff = 8., years = 10., Probability= 5.93677*10-11 per quake or 3.21631*10-12 per day, count = 0.001, sd = 0.0104429, mean number of quakes in time period is: 178.086, Number of days in period = 3652.42

magnitude cutoff = 8., years = 20., Probability= 6.53153*10-11 per quake or 3.37712*10-12 per day, count = 0.0021, sd = 0.0150234, mean number of quakes in time period is: 339.927, Number of days in period = 7304.84

magnitude cutoff = 8., years = 30., Probability= 6.2131*10-11 per quake or 3.1627*10-12 per day, count = 0.00295, sd = 0.0175057, mean number of quakes in time period is: 501.989, Number of days in period = 10957.3

Chile – Argentina

magnitude cutoff = 6., years = 0.00436183, Probability= 1.60242*10-6 per quake or 1.90957*10-6 per day, count = 1.0595, sd = 0.966812, mean number of quakes in time period is: 0.210944, Number of days in period = 1.59312

magnitude cutoff = 6., years = 10., Probability= 1.48883*10-6 per quake or 8.14048*10-7 per day, count = 869.61, sd = 186.967, mean number of quakes in time period is: 1797.33, Number of days in period = 3652.42

magnitude cutoff = 6., years = 20., Probability= 1.51649*10-6 per quake or 9.20385*10-7 per day, count = 1965.56, sd = 384.468, mean number of quakes in time period is: 3990.08, Number of days in period = 7304.84

magnitude cutoff = 6., years = 30., Probability= 1.50408*10-6 per quake or 1.00293*10-6 per day, count = 3212.29, sd = 601.546, mean number of quakes in time period is: 6575.75, Number of days in period = 10957.3

magnitude cutoff = 6.5, years = 0.00591204, Probability= 1.07347*10-6 per quake or 1.22419*10-6 per day, count = 1.098, sd = 1.00524, mean number of quakes in time period is: 0.273611, Number of days in period = 2.15933

magnitude cutoff = 6.5, years = 10., Probability= 8.35226*10-7 per quake or 4.56677*10-7 per day, count = 609.225, sd = 138.16, mean number of quakes in time period is: 1797.33, Number of days in period = 3652.42

magnitude cutoff = 6.5, years = 20., Probability= 8.50409*10-7 per quake or 5.16127*10-7 per day, count = 1376.13, sd = 276.041, mean number of quakes in time period is: 3990.08, Number of days in period = 7304.84

magnitude cutoff = 6.5, years = 30., Probability= 8.43344*10-7 per quake or 5.62348*10-7 per day, count = 2248.8, sd = 427.692, mean number of quakes in time period is: 6575.75, Number of days in period = 10957.3

magnitude cutoff = 7., years = 0.00832344, Probability= 7.18654*10-7 per quake or 7.77853*10-7 per day, count = 1.0665, sd = 0.981511, mean number of quakes in time period is: 0.365611, Number of days in period = 3.04007

magnitude cutoff = 7., years = 10., Probability= 4.76082*10-7 per quake or 2.60308*10-7 per day, count = 381.059, sd = 90.0593, mean number of quakes in time period is: 1797.33, Number of days in period = 3652.42

magnitude cutoff = 7., years = 20., Probability= 4.82717*10-7 per quake or 2.92969*10-7 per day, count = 857.562, sd = 175.517, mean number of quakes in time period is: 3990.08, Number of days in period = 7304.84

magnitude cutoff = 7., years = 30., Probability= 4.78401*10-7 per quake or 3.19001*10-7 per day, count = 1400.56, sd = 269.933, mean number of quakes in time period is: 6575.75, Number of days in period = 10957.3

magnitude cutoff = 7.5, years = 0.013837, Probability= 4.39961*10-7 per quake or 4.42932*10-7 per day, count = 1.0235, sd = 0.952171, mean number of quakes in time period is: 0.565333, Number of days in period = 5.05387

magnitude cutoff = 7.5, years = 10., Probability= 2.38153*10-7 per quake or 1.30215*10-7 per day, count = 192.694, sd = 47.0539, mean number of quakes in time period is: 1797.33, Number of days in period = 3652.42

magnitude cutoff = 7.5, years = 20., Probability= 2.3994*10-7 per quake or 1.45624*10-7 per day, count = 430.889, sd = 89.7149, mean number of quakes in time period is: 3990.08, Number of days in period = 7304.84

magnitude cutoff = 7.5, years = 30., Probability= 2.37391*10-7 per quake or 1.58294*10-7 per day, count = 702.515, sd = 136.979, mean number of quakes in time period is: 6575.75, Number of days in period = 10957.3

magnitude cutoff = 8., years = 0.0282712, Probability= 2.42486*10-7 per quake or 2.2294*10-7 per day, count = 1.0725, sd = 1.01424, mean number of quakes in time period is: 1.05483, Number of days in period = 10.3258

magnitude cutoff = 8., years = 10., Probability= 9.59456*10-8 per quake or 5.24602*10-8 per day, count = 79.7859, sd = 19.8915, mean number of quakes in time period is: 1797.33, Number of days in period = 3652.42

magnitude cutoff = 8., years = 20., Probability= 9.60119*10-8 per quake or 5.82711*10-8 per day, count = 177.207, sd = 37.2839, mean number of quakes in time period is: 3990.08, Number of days in period = 7304.84

magnitude cutoff = 8., years = 30., Probability= 9.47932*10-8 per quake or 6.32088*10-8 per day, count = 288.313, sd = 56.6051, mean number of quakes in time period is: 6575.75, Number of days in period = 10957.3

magnitude cutoff = 8.5, years = 0.335395, Probability= 2.56332*10-8 per quake or 1.69686*10-8 per day, count = 0.9865, sd = 0.910744, mean number of quakes in time period is: 9.01028, Number of days in period = 122.5

magnitude cutoff = 8.5, years = 10., Probability= 1.40237*10-8 per quake or 7.66775*10-9 per day, count = 11.9621, sd = 3.02283, mean number of quakes in time period is: 1797.33, Number of days in period = 3652.42

magnitude cutoff = 8.5, years = 20., Probability= 1.38649*10-8 per quake or 8.41482*10-9 per day, count = 26.2551, sd = 5.60224, mean number of quakes in time period is: 3990.08, Number of days in period = 7304.84

magnitude cutoff = 8.5, years = 30., Probability= 1.36373*10-8 per quake or 9.09347*10-9 per day, count = 42.5588, sd = 8.45008, mean number of quakes in time period is: 6575.75, Number of days in period = 10957.3

magnitude cutoff = 9., years = 5.17282, Probability= 2.38104*10-9 per quake or 1.13032*10-9 per day, count = 1.0135, sd = 0.920089, mean number of quakes in time period is: 99.6553, Number of days in period = 1889.33

magnitude cutoff = 9., years = 10., Probability= 2.27618*10-9 per quake or 1.24455*10-9 per day, count = 1.94155, sd = 0.552867, mean number of quakes in time period is: 1797.33, Number of days in period = 3652.42

magnitude cutoff = 9., years = 20., Probability= 2.25318*10-9 per quake or 1.36749*10-9 per day, count = 4.2667, sd = 0.992947, mean number of quakes in time period is: 3990.08, Number of days in period = 7304.84

magnitude cutoff = 9., years = 30., Probability= 2.21719*10-9 per quake or 1.47844*10-9 per day, count = 6.9193, sd = 1.47228, mean number of quakes in time period is: 6575.75, Number of days in period = 10957.3

South Africa

magnitude cutoff = 5., years = 0.0528589, Probability= 3.76155*10-6 per quake or 6.34384*10-7 per day, count = 0.954, sd = 0.880665, mean number of quakes in time period is: 0.361778, Number of days in period = 19.3063

magnitude cutoff = 5., years = 10., Probability= 2.96949*10-6 per quake or 1.80644*10-7 per day, count = 42.5004, sd = 11.1146, mean number of quakes in time period is: 199.97, Number of days in period = 3652.42

magnitude cutoff = 5., years = 20., Probability= 2.95726*10-6 per quake or 1.72185*10-7 per day, count = 80.836, sd = 21.3989, mean number of quakes in time period is: 382.79, Number of days in period = 7304.84

magnitude cutoff = 5., years = 30., Probability= 2.94616*10-6 per quake or 1.75551*10-7 per day, count = 123.586, sd = 33.0297, mean number of quakes in time period is: 587.612, Number of days in period = 10957.3

magnitude cutoff = 5.5, years = 0.146527, Probability= 1.68778*10-6 per quake or 2.01172*10-7 per day, count = 1.03, sd = 0.950127, mean number of quakes in time period is: 0.708778, Number of days in period = 53.518

magnitude cutoff = 5.5, years = 10., Probability= 1.06161*10-6 per quake or 6.45811*10-8 per day, count = 20.5447, sd = 5.48931, mean number of quakes in time period is: 199.97, Number of days in period = 3652.42

magnitude cutoff = 5.5, years = 20., Probability= 1.04449*10-6 per quake or 6.0815*10-8 per day, count = 38.7597, sd = 10.362, mean number of quakes in time period is: 382.79, Number of days in period = 7304.84

magnitude cutoff = 5.5, years = 30., Probability= 1.03725*10-6 per quake or 6.1806*10-8 per day, count = 59.1268, sd = 15.9136, mean number of quakes in time period is: 587.612, Number of days in period = 10957.3

magnitude cutoff = 6., years = 0.721929, Probability= 3.96169*10-7 per quake or 3.43374*10-8 per day, count = 0.994, sd = 0.906774, mean number of quakes in time period is: 2.53933, Number of days in period = 263.679

magnitude cutoff = 6., years = 10., Probability= 2.77928*10-7 per quake or 1.69073*10-8 per day, count = 6.0751, sd = 1.66435, mean number of quakes in time period is: 199.97, Number of days in period = 3652.42

magnitude cutoff = 6., years = 20., Probability= 2.7286*10-7 per quake or 1.58872*10-8 per day, count = 11.4, sd = 3.09548, mean number of quakes in time period is: 382.79, Number of days in period = 7304.84

magnitude cutoff = 6., years = 30., Probability= 2.70807*10-7 per quake or 1.61363*10-8 per day, count = 17.3532, sd = 4.72377, mean number of quakes in time period is: 587.612, Number of days in period = 10957.3

magnitude cutoff = 6.5, years = 7.90962, Probability= 5.22371*10-8 per quake or 2.92926*10-9 per day, count = 0.983, sd = 0.893146, mean number of quakes in time period is: 18., Number of days in period = 2888.93

magnitude cutoff = 6.5, years = 10., Probability= 5.15074*10-8 per quake or 3.13337*10-9 per day, count = 1.19645, sd = 0.431921, mean number of quakes in time period is: 199.97, Number of days in period = 3652.42

magnitude cutoff = 6.5, years = 20., Probability= 5.12615*10-8 per quake or 2.98468*10-9 per day, count = 2.27935, sd = 0.723637, mean number of quakes in time period is: 382.79, Number of days in period = 7304.84

magnitude cutoff = 6.5, years = 30., Probability= 5.11783*10-8 per quake or 3.04952*10-9 per day, count = 3.4933, sd = 1.07046, mean number of quakes in time period is: 587.612, Number of days in period = 10957.3

magnitude cutoff = 7., years = 30., Probability= 1.68481*10-10 per quake or 9.03516*10-12 per day, count = 0.0115, sd = 0.0148631, mean number of quakes in time period is: 65.2899, Number of days in period = 10957.3

magnitude cutoff = 7., years = 10., Probability= 1.76506*10-10 per quake or 1.07374*10-11 per day, count = 0.0041, sd = 0.019834, mean number of quakes in time period is: 199.97, Number of days in period = 3652.42

magnitude cutoff = 7., years = 20., Probability= 1.63049*10-10 per quake or 9.49347*10-12 per day, count = 0.00725, sd = 0.0265102, mean number of quakes in time period is: 382.79, Number of days in period = 7304.84

magnitude cutoff = 7., years = 30., Probability= 1.53097*10-10 per quake or 9.12246*10-12 per day, count = 0.01045, sd = 0.0321919, mean number of quakes in time period is: 587.612, Number of days in period = 10957.3

magnitude cutoff = 7.5, years = 30., Probability= 0. per quake or 0. per day, count = 0., sd = 0., mean number of quakes in time period is: 65.2899, Number of days in period = 10957.3

magnitude cutoff = 7.5, years = 10., Probability= 0. per quake or 0. per day, count = 0., sd = 0., mean number of quakes in time period is: 199.97, Number of days in period = 3652.42

magnitude cutoff = 7.5, years = 20., Probability= 0. per quake or 0. per day, count = 0., sd = 0., mean number of quakes in time period is: 382.79, Number of days in period = 7304.84

magnitude cutoff = 7.5, years = 30., Probability= 0. per quake or 0. per day, count = 0., sd = 0., mean number of quakes in time period is: 587.612, Number of days in period = 10957.3

Country = Tibet-Nepal-Bhutan

magnitude cutoff = 5., years = 0.0191431, Probability= 6.05742*10-6 per quake or 1.53128*10-6 per day, count = 1.001, sd = 0.914209, mean number of quakes in time period is: 0.196389, Number of days in period = 6.99187

magnitude cutoff = 5., years = 10., Probability= 6.52281*10-6 per quake or 6.59222*10-7 per day, count = 178.962, sd = 28.2624, mean number of quakes in time period is: 332.216, Number of days in period = 3652.42

magnitude cutoff = 5., years = 20., Probability= 6.6297*10-6 per quake or 6.64226*10-7 per day, count = 360.386, sd = 48.9302, mean number of quakes in time period is: 658.681, Number of days in period = 7304.84

magnitude cutoff = 5., years = 30., Probability= 6.65932*10-6 per quake or 6.74019*10-7 per day, count = 548.639, sd = 74.0683, mean number of quakes in time period is: 998.13, Number of days in period = 10957.3

magnitude cutoff = 5.5, years = 0.0263377, Probability= 3.61923*10-6 per quake or 8.40505*10-7 per day, count = 1.004, sd = 0.919225, mean number of quakes in time period is: 0.248222, Number of days in period = 9.61964

magnitude cutoff = 5.5, years = 10., Probability= 3.2899*10-6 per quake or 3.32491*10-7 per day, count = 129.835, sd = 22.3202, mean number of quakes in time period is: 332.216, Number of days in period = 3652.42

magnitude cutoff = 5.5, years = 20., Probability= 3.3437*10-6 per quake or 3.35004*10-7 per day, count = 261.347, sd = 37.4189, mean number of quakes in time period is: 658.681, Number of days in period = 7304.84

magnitude cutoff = 5.5, years = 30., Probability= 3.36073*10-6 per quake or 3.40154*10-7 per day, count = 397.958, sd = 55.4469, mean number of quakes in time period is: 998.13, Number of days in period = 10957.3

magnitude cutoff = 6., years = 0.0416562, Probability= 2.33392*10-6 per quake or 4.72319*10-7 per day, count = 1., sd = 0.919877, mean number of quakes in time period is: 0.342111, Number of days in period = 15.2146

magnitude cutoff = 6., years = 10., Probability= 1.89733*10-6 per quake or 1.91752*10-7 per day, count = 86.1368, sd = 15.7262, mean number of quakes in time period is: 332.216, Number of days in period = 3652.42

magnitude cutoff = 6., years = 20., Probability= 1.92384*10-6 per quake or 1.92749*10-7 per day, count = 172.976, sd = 25.9957, mean number of quakes in time period is: 658.681, Number of days in period = 7304.84

magnitude cutoff = 6., years = 30., Probability= 1.93273*10-6 per quake or 1.95621*10-7 per day, count = 263.257, sd = 37.8635, mean number of quakes in time period is: 998.13, Number of days in period = 10957.3

magnitude cutoff = 6.5, years = 0.0738444, Probability= 1.31486*10-6 per quake or 2.32321*10-7 per day, count = 0.982, sd = 0.908648, mean number of quakes in time period is: 0.5295, Number of days in period = 26.9711

magnitude cutoff = 6.5, years = 10., Probability= 9.4709*10-7 per quake or 9.57169*10-8 per day, count = 49.1343, sd = 9.43322, mean number of quakes in time period is: 332.216, Number of days in period = 3652.42

magnitude cutoff = 6.5, years = 20., Probability= 9.56246*10-7 per quake or 9.58058*10-8 per day, count = 98.3093, sd = 15.3276, mean number of quakes in time period is: 658.681, Number of days in period = 7304.84

magnitude cutoff = 6.5, years = 30., Probability= 9.59722*10-7 per quake or 9.71377*10-8 per day, count = 149.483, sd = 22.0859, mean number of quakes in time period is: 998.13, Number of days in period = 10957.3

magnitude cutoff = 7., years = 0.189081, Probability= 6.04918*10-7 per quake or 8.92349*10-8 per day, count = 0.987, sd = 0.913517, mean number of quakes in time period is: 1.13194, Number of days in period = 69.0604

magnitude cutoff = 7., years = 10., Probability= 4.04445*10-7 per quake or 4.08749*10-8 per day, count = 21.5196, sd = 4.29293, mean number of quakes in time period is: 332.216, Number of days in period = 3652.42

magnitude cutoff = 7., years = 20., Probability= 4.06568*10-7 per quake or 4.07338*10-8 per day, count = 42.8905, sd = 6.88991, mean number of quakes in time period is: 658.681, Number of days in period = 7304.84

magnitude cutoff = 7., years = 30., Probability= 4.07623*10-7 per quake or 4.12573*10-8 per day, count = 65.1627, sd = 9.81106, mean number of quakes in time period is: 998.13, Number of days in period = 10957.3

magnitude cutoff = 7.5, years = 0.996711, Probability= 1.47211*10-7 per quake or 1.70141*10-8 per day, count = 0.992, sd = 0.904207, mean number of quakes in time period is: 4.67494, Number of days in period = 364.041

magnitude cutoff = 7.5, years = 10., Probability= 1.20682*10-7 per quake or 1.21966*10-8 per day, count = 6.4212, sd = 1.33948, mean number of quakes in time period is: 332.216, Number of days in period = 3652.42

magnitude cutoff = 7.5, years = 20., Probability= 1.21308*10-7 per quake or 1.21538*10-8 per day, count = 12.7973, sd = 2.11659, mean number of quakes in time period is: 658.681, Number of days in period = 7304.84

magnitude cutoff = 7.5, years = 30., Probability= 1.21589*10-7 per quake or 1.23066*10-8 per day, count = 19.4373, sd = 2.9856, mean number of quakes in time period is: 998.13, Number of days in period = 10957.3

magnitude cutoff = 8., years = 30., Probability= 0. per quake or 0. per day, count = 0., sd = 0., mean number of quakes in time period is: 111.437, Number of days in period = 10957.3

magnitude cutoff = 8., years = 10., Probability= 0. per quake or 0. per day, count = 0., sd = 0., mean number of quakes in time period is: 332.216, Number of days in period = 3652.42

magnitude cutoff = 8., years = 20., Probability= 0. per quake or 0. per day, count = 0., sd = 0., mean number of quakes in time period is: 658.681, Number of days in period = 7304.84

magnitude cutoff = 8., years = 30., Probability= 0. per quake or 0. per day, count = 0., sd = 0., mean number of quakes in time period is: 998.13, Number of days in period = 10957.3

Country = Australia

magnitude cutoff = 5., years = 0.128373, Probability= 0.0000112395 per quake or 5.21378*10-7 per day, count = 1.044, sd = 0.952435, mean number of quakes in time period is: 0.241667, Number of days in period = 46.8872

magnitude cutoff = 5., years = 10., Probability= 0.0000110405 per quake or 3.89495*10-7 per day, count = 52.9651, sd = 6.42154, mean number of quakes in time period is: 115.968, Number of days in period = 3652.42

magnitude cutoff = 5., years = 20., Probability= 0.0000111616 per quake or 3.84989*10-7 per day, count = 104.51, sd = 10.3277, mean number of quakes in time period is: 226.765, Number of days in period = 7304.84

magnitude cutoff = 5., years = 30., Probability= 0.0000112142 per quake or 3.83619*10-7 per day, count = 156.146, sd = 14.4926, mean number of quakes in time period is: 337.346, Number of days in period = 10957.3

magnitude cutoff = 5.5, years = 0.24447, Probability= 4.06654*10-6 per quake or 1.69646*10-7 per day, count = 1.004, sd = 0.920237, mean number of quakes in time period is: 0.413889, Number of days in period = 89.2908

magnitude cutoff = 5.5, years = 10., Probability= 3.73172*10-6 per quake or 1.31651*10-7 per day, count = 28.6128, sd = 3.89717, mean number of quakes in time period is: 115.968, Number of days in period = 3652.42

magnitude cutoff = 5.5, years = 20., Probability= 3.75935*10-6 per quake or 1.29669*10-7 per day, count = 56.3415, sd = 6.08322, mean number of quakes in time period is: 226.765, Number of days in period = 7304.84

magnitude cutoff = 5.5, years = 30., Probability= 3.77338*10-6 per quake or 1.29081*10-7 per day, count = 84.1204, sd = 8.36997, mean number of quakes in time period is: 337.346, Number of days in period = 10957.3

magnitude cutoff = 6., years = 0.8314, Probability= 1.29221*10-6 per quake or 4.5967*10-8 per day, count = 1.0105, sd = 0.92052, mean number of quakes in time period is: 1.20022, Number of days in period = 303.662

magnitude cutoff = 6., years = 10., Probability= 1.1439*10-6 per quake or 4.03556*10-8 per day, count = 9.63405, sd = 1.42959, mean number of quakes in time period is: 115.968, Number of days in period = 3652.42

magnitude cutoff = 6., years = 20., Probability= 1.14635*10-6 per quake or 3.95404*10-8 per day, count = 18.8899, sd = 2.18994, mean number of quakes in time period is: 226.765, Number of days in period = 7304.84

magnitude cutoff = 6., years = 30., Probability= 1.14938*10-6 per quake or 3.93183*10-8 per day, count = 28.1676, sd = 2.95969, mean number of quakes in time period is: 337.346, Number of days in period = 10957.3

magnitude cutoff = 6.5, years = 5.21835, Probability= 1.92694*10-7 per quake or 6.24864*10-9 per day, count = 1.0005, sd = 0.90675, mean number of quakes in time period is: 6.86733, Number of days in period = 1905.96

magnitude cutoff = 6.5, years = 10., Probability= 1.92206*10-7 per quake or 6.7808*10-9 per day, count = 1.8725, sd = 0.452389, mean number of quakes in time period is: 115.968, Number of days in period = 3652.42

magnitude cutoff = 6.5, years = 20., Probability= 1.92439*10-7 per quake or 6.63767*10-9 per day, count = 3.66595, sd = 0.658812, mean number of quakes in time period is: 226.765, Number of days in period = 7304.84

magnitude cutoff = 6.5, years = 30., Probability= 1.92826*10-7 per quake or 6.59624*10-9 per day, count = 5.4646, sd = 0.842408, mean number of quakes in time period is: 337.346, Number of days in period = 10957.3

magnitude cutoff = 7., years = 30., Probability= 0. per quake or 0. per day, count = 0., sd = 0., mean number of quakes in time period is: 37.3599, Number of days in period = 10957.3

Country = Cuba

magnitude cutoff = 5., years = 0.521646, Probability= 0.000015549 per quake or 2.27857*10-7 per day, count = 1.0395, sd = 0.948706, mean number of quakes in time period is: 0.310222, Number of days in period = 190.527

magnitude cutoff = 5., years = 10., Probability= 0.0000164077 per quake or 1.23618*10-7 per day, count = 9.6909, sd = 2.06091, mean number of quakes in time period is: 24.7661, Number of days in period = 3652.42

magnitude cutoff = 5., years = 20., Probability= 0.0000165679 per quake or 1.07178*10-7 per day, count = 16.8148, sd = 3.24681, mean number of quakes in time period is: 42.5295, Number of days in period = 7304.84

magnitude cutoff = 5., years = 30., Probability= 0.0000166607 per quake or 1.00968*10-7 per day, count = 23.766, sd = 4.11106, mean number of quakes in time period is: 59.7635, Number of days in period = 10957.

Country = Brazil

magnitude cutoff = 4.5, years = 0.0523135, Probability= 0.0000167877 per quake or 1.35525*10^-6 per day, count = 1.026, sd = 0.933551, mean number of quakes in time period is: 0.171389, Number of days in period = 19.1071

magnitude cutoff = 4.5, years = 10., Probability= 0.0000193468 per quake or 9.11348*10^-7 per day, count = 101.695, sd = 11.4168, mean number of quakes in time period is: 154.846, Number of days in period = 3652.42

magnitude cutoff = 4.5, years = 20., Probability= 0.0000195591 per quake or 9.0545*10^-7 per day, count = 201.916, sd = 15.7555, mean number of quakes in time period is: 304.348, Number of days in period = 7304.84

magnitude cutoff = 4.5, years = 30., Probability= 0.0000196507 per quake or 9.05494*10^-7 per day, count = 303.043, sd = 19.4632, mean number of quakes in time period is: 454.415, Number of days in period = 10957.3

magnitude cutoff = 5., years = 0.0673053, Probability= 8.57066*10^-6 per quake or 6.63472*10^-7 per day, count = 0.9905, sd = 0.903446, mean number of quakes in time period is: 0.211444, Number of days in period = 24.5827

magnitude cutoff = 5., years = 10., Probability= 8.51817*10^-6 per quake or 4.01256*10^-7 per day, count = 74.8633, sd = 9.64522, mean number of quakes in time period is: 154.846, Number of days in period = 3652.42

magnitude cutoff = 5., years = 20., Probability= 8.61312*10^-6 per quake or 3.98728*10^-7 per day, count = 148.579, sd = 13.2148, mean number of quakes in time period is: 304.348, Number of days in period = 7304.84

magnitude cutoff = 5., years = 30., Probability= 8.66968*10^-6 per quake or 3.99495*10^-7 per day, count = 223.1, sd = 16.2369, mean number of quakes in time period is: 454.415, Number of days in period = 10957.3

magnitude cutoff = 5.5, years = 0.0961532, Probability= 4.89327*10^-6 per quake or 3.54673*10^-7 per day, count = 0.9775, sd = 0.894746, mean number of quakes in time period is: 0.282833, Number of days in period = 35.1192

magnitude cutoff = 5.5, years = 10., Probability= 4.23861*10^-6 per quake or 1.99664*10^-7 per day, count = 51.0932, sd = 7.23021, mean number of quakes in time period is: 154.846, Number of days in period = 3652.42

magnitude cutoff = 5.5, years = 20., Probability= 4.27234*10^-6 per quake or 1.9778*10^-7 per day, count = 101.205, sd = 9.91004, mean number of quakes in time period is: 304.348, Number of days in period = 7304.84

magnitude cutoff = 5.5, years = 30., Probability= 4.29245*10^-6 per quake or 1.97794*10^-7 per day, count = 151.803, sd = 12.115, mean number of quakes in time period is: 454.415, Number of days in period = 10957.3

magnitude cutoff = 6., years = 0.154364, Probability= 2.90819*10^-6 per quake or 1.97095*10^-7 per day, count = 0.9355, sd = 0.859629, mean number of quakes in time period is: 0.424556, Number of days in period = 56.3801

magnitude cutoff = 6., years = 10., Probability= 2.40584*10^-6 per quake or 1.13329*10^-7 per day, count = 31.413, sd = 4.7297, mean number of quakes in time period is: 154.846, Number of days in period = 3652.42

magnitude cutoff = 6., years = 20., Probability= 2.41765*10^-6 per quake or 1.1192*10^-7 per day, count = 62.0412, sd = 6.49453, mean number of quakes in time period is: 304.348, Number of days in period = 7304.84

magnitude cutoff = 6., years = 30., Probability= 2.42676*10^-6 per quake or 1.11824*10^-7 per day, count = 92.9873, sd = 7.96097, mean number of quakes in time period is: 454.415, Number of days in period = 10957.3

magnitude cutoff = 6.5, years = 0.351219, Probability= 1.46268*10^-6 per quake or 8.82931*10^-8 per day, count = 0.934, sd = 0.856295, mean number of quakes in time period is: 0.860389, Number of days in period = 128.28

magnitude cutoff = 6.5, years = 10., Probability= 1.20696*10^-6 per quake or 5.68547*10^-8 per day, count = 15.4197, sd = 2.4359, mean number of quakes in time period is: 154.846, Number of days in period = 3652.42

magnitude cutoff = 6.5, years = 20., Probability= 1.20941*10^-6 per quake or 5.59875*10^-8 per day, count = 30.3616, sd = 3.35566, mean number of quakes in time period is: 304.348, Number of days in period = 7304.84

magnitude cutoff = 6.5, years = 30., Probability= 1.21321*10^-6 per quake or 5.59041*10^-8 per day, count = 45.4792, sd = 4.13828, mean number of quakes in time period is: 454.415, Number of days in period = 10957.3

magnitude cutoff = 7., years = 1.94104, Probability= 3.27806*10^-7 per quake or 1.56627*10^-8 per day, count = 0.9555, sd = 0.867038, mean number of quakes in time period is: 3.76378, Number of days in period = 708.95

magnitude cutoff = 7., years = 10., Probability= 3.11017*10^-7 per quake or 1.46508*10^-8 per day, count = 4.1441, sd = 0.726842, mean number of quakes in time period is: 154.846, Number of days in period = 3652.42

magnitude cutoff = 7., years = 20., Probability= 3.09752*10^-7 per quake or 1.43394*10^-8 per day, count = 8.11205, sd = 1.01138, mean number of quakes in time period is: 304.348, Number of days in period = 7304.84

magnitude cutoff = 7., years = 30., Probability= 3.10104*10^-7 per quake or 1.42894*10^-8 per day, count = 12.1257, sd = 1.24694, mean number of quakes in time period is: 454.415, Number of days in period = 10957.3

magnitude cutoff = 7.5, years = 30., Probability= 0. per quake or 0. per day, count = 0., sd = 0., mean number of quakes in time period is: 50.4289, Number of days in period = 10957.3

Country = Israel-Syria

magnitude cutoff = 4.5, years = 0.474382, Probability= 0.000017743 per quake or 3.0414*10^-7 per day, count = 1.054, sd = 0.960163, mean number of quakes in time period is: 0.33, Number of days in period = 173.264

magnitude cutoff = 4.5, years = 10., Probability= 0.0000180597 per quake or 1.5122*10^-7 per day, count = 9.9036, sd = 2.05861, mean number of quakes in time period is: 27.5247, Number of days in period = 3652.42

magnitude cutoff = 4.5, years = 20., Probability= 0.0000182325 per quake or 1.29663*10^-7 per day, count = 16.9646, sd = 3.2892, mean number of quakes in time period is: 46.7544, Number of days in period = 7304.84

magnitude cutoff = 4.5, years = 30., Probability= 0.0000182577 per quake or 1.18281*10^-7 per day, count = 23.2196, sd = 4.30265, mean number of quakes in time period is: 63.8871, Number of days in period = 10957.3

magnitude cutoff = 5., years = 1.4204, Probability= 7.59599*10^-6 per quake or 9.20965*10^-8 per day, count = 1.037, sd = 0.941269, mean number of quakes in time period is: 0.698889, Number of days in period = 518.79

magnitude cutoff = 5., years = 10., Probability= 7.50999*10^-6 per quake or 6.28836*10^-8 per day, count = 4.5006, sd = 0.976302, mean number of quakes in time period is: 27.5247, Number of days in period = 3652.42

magnitude cutoff = 5., years = 20., Probability= 7.57472*10^-6 per quake or 5.38685*10^-8 per day, count = 7.69885, sd = 1.5341, mean number of quakes in time period is: 46.7544, Number of days in period = 7304.84

magnitude cutoff = 5., years = 30., Probability= 7.579*10^-6 per quake or 4.90999*10^-8 per day, count = 10.5308, sd = 1.9963, mean number of quakes in time period is: 63.8871, Number of days in period = 10957.3

magnitude cutoff = 5.5, years = 5.32568, Probability= 2.66484*10^-6 per quake or 2.27582*10^-8 per day, count = 1.015, sd = 0.921083, mean number of quakes in time period is: 1.84578, Number of days in period = 1945.16

magnitude cutoff = 5.5, years = 10., Probability= 2.65047*10^-6 per quake or 2.21933*10^-8 per day, count = 1.6567, sd = 0.469814, mean number of quakes in time period is: 27.5247, Number of days in period = 3652.42

magnitude cutoff = 5.5, years = 20., Probability= 2.67894*10^-6 per quake or 1.90516*10^-8 per day, count = 2.83545, sd = 0.689032, mean number of quakes in time period is: 46.7544, Number of days in period = 7304.84

magnitude cutoff = 5.5, years = 30., Probability= 2.67958*10^-6 per quake or 1.73594*10^-8 per day, count = 3.879, sd = 0.881429, mean number of quakes in time period is: 63.8871, Number of days in period = 10957.3

magnitude cutoff = 6., years = 30., Probability= 0. per quake or 0. per day, count = 0., sd = 0., mean number of quakes in time period is: 7.12706, Number of days in period = 10957.3

Country = France-Monaco

magnitude cutoff = 4.5, years = 0.00763878, Probability= 8.46875*10^-7 per quake or 8.77228*10^-7 per day, count = 0.965, sd = 0.875038, mean number of quakes in time period is: 0.412857, Number of days in period = 2.79

magnitude cutoff = 4.5, years = 10., Probability= 6.74981*10^-7 per quake or 4.50754*10^-7 per day, count = 559.078, sd = 116.026, mean number of quakes in time period is: 2134.21, Number of days in period = 3652.42

magnitude cutoff = 4.5, years = 20., Probability= 6.88606*10^-7 per quake or 4.88674*10^-7 per day, count = 1211.81, sd = 233.177, mean number of quakes in time period is: 4535.95, Number of days in period = 7304.84

magnitude cutoff = 4.5, years = 30., Probability= 6.9322*10^-7 per quake or 5.26643*10^-7 per day, count = 1958.7, sd = 369.35, mean number of quakes in time period is: 7283.77, Number of days in period = 10957.3

magnitude cutoff = 5., years = 0.0180269, Probability= 3.92647*10^-7 per quake or 3.71585*10^-7 per day, count = 1.0145, sd = 0.944368, mean number of quakes in time period is: 0.890143, Number of days in period = 6.58419

magnitude cutoff = 5., years = 10., Probability= 2.23963*10^-7 per quake or 1.49563*10^-7 per day, count = 197.404, sd = 43.2932, mean number of quakes in time period is: 2134.21, Number of days in period = 3652.42

magnitude cutoff = 5., years = 20., Probability= 2.27388*10^-7 per quake or 1.61368*10^-7 per day, count = 425.887, sd = 84.4664, mean number of quakes in time period is: 4535.95, Number of days in period = 7304.84

magnitude cutoff = 5., years = 30., Probability= 2.28381*10^-7 per quake or 1.73502*10^-7 per day, count = 686.838, sd = 131.645, mean number of quakes in time period is: 7283.77, Number of days in period = 10957.3

magnitude cutoff = 5.5, years = 0.675009, Probability= 1.48602*10^-8 per quake or 9.61386*10^-9 per day, count = 1.0125, sd = 0.908092, mean number of quakes in time period is: 22.7859, Number of days in period = 246.542

magnitude cutoff = 5.5, years = 10., Probability= 1.05928*10^-8 per quake or 7.07392*10^-9 per day, count = 9.65731, sd = 2.19814, mean number of quakes in time period is: 2134.21, Number of days in period = 3652.42

magnitude cutoff = 5.5, years = 20., Probability= 1.06521*10^-8 per quake or 7.55933*10^-9 per day, count = 20.64, sd = 4.19629, mean number of quakes in time period is: 4535.95, Number of days in period = 7304.84

magnitude cutoff = 5.5, years = 30., Probability= 1.06666*10^-8 per quake or 8.10345*10^-9 per day, count = 33.1885, sd = 6.4746, mean number of quakes in time period is: 7283.77, Number of days in period = 10957.3

magnitude cutoff = 6., years = 30., Probability= 0. per quake or 0. per day, count = 0., sd = 0., mean number of quakes in time period is: 1039.87, Number of days in period = 10957.3

Country = Egypt-Libya

magnitude cutoff = 4.5, years = 0.132935, Probability= 0.0000108535 per quake or 9.56181*10^-7 per day, count = 1.933, sd = 1.78398, mean number of quakes in time period is: 4.2775, Number of days in period = 48.5533

magnitude cutoff = 4.5, years = 10., Probability= 0.0000100145 per quake or 6.64798*10^-7 per day, count = 50.2215, sd = 18.3953, mean number of quakes in time period is: 121.231, Number of days in period = 3652.42

magnitude cutoff = 4.5, years = 20., Probability= 0.0000100992 per quake or 6.38893*10^-7 per day, count = 96.4178, sd = 25.048, mean number of quakes in time period is: 231.059, Number of days in period = 7304.84

magnitude cutoff = 4.5, years = 30., Probability= 0.0000101401 per quake or 6.29982*10^-7 per day, count = 142.592, sd = 31.1273, mean number of quakes in time period is: 340.376, Number of days in period = 10957.3

magnitude cutoff = 5., years = 0.213174, Probability= 6.42409*10^-6 per quake or 4.59198*10^-7 per day, count = 1.66, sd = 1.54449, mean number of quakes in time period is: 5.5655, Number of days in period = 77.8602

magnitude cutoff = 5., years = 10., Probability= 5.49928*10^-6 per quake or 3.65064*10^-7 per day, count = 30.962, sd = 12.0464, mean number of quakes in time period is: 121.231, Number of days in period = 3652.42

magnitude cutoff = 5., years = 20., Probability= 5.53306*10^-6 per quake or 3.50032*10^-7 per day, count = 59.2988, sd = 16.3929, mean number of quakes in time period is: 231.059, Number of days in period = 7304.84

magnitude cutoff = 5., years = 30., Probability= 5.55138*10^-6 per quake or 3.44895*10^-7 per day, count = 87.645, sd = 20.0656, mean number of quakes in time period is: 340.376, Number of days in period = 10957.3

magnitude cutoff = 5.5, years = 0.398168, Probability= 3.14099*10^-6 per quake or 1.8054*10^-7 per day, count = 1.3475, sd = 1.19377, mean number of quakes in time period is: 8.359, Number of days in period = 145.428

magnitude cutoff = 5.5, years = 10., Probability= 2.60691*10^-6 per quake or 1.73057*10^-7 per day, count = 16.2185, sd = 6.56489, mean number of quakes in time period is: 121.231, Number of days in period = 3652.42

magnitude cutoff = 5.5, years = 20., Probability= 2.61435*10^-6 per quake or 1.65389*10^-7 per day, count = 31.0083, sd = 8.94738, mean number of quakes in time period is: 231.059, Number of days in period = 7304.84

magnitude cutoff = 5.5, years = 30., Probability= 2.62261*10^-6 per quake or 1.62937*10^-7 per day, count = 45.7928, sd = 10.8531, mean number of quakes in time period is: 340.376, Number of days in period = 10957.3

magnitude cutoff = 6., years = 0.978953, Probability= 1.06088*10^-6 per quake or 5.07036*10^-8 per day, count = 1.016, sd = 0.795812, mean number of quakes in time period is: 17.089, Number of days in period = 357.555

magnitude cutoff = 6., years = 10., Probability= 9.58249*10^-7 per quake or 6.36123*10^-8 per day, count = 6.4855, sd = 2.70304, mean number of quakes in time period is: 121.231, Number of days in period = 3652.42

magnitude cutoff = 6., years = 20., Probability= 9.5791*10^-7 per quake or 6.05992*10^-8 per day, count = 12.369, sd = 3.70796, mean number of quakes in time period is: 231.059, Number of days in period = 7304.84

magnitude cutoff = 6., years = 30., Probability= 9.58849*10^-7 per quake or 5.95712*10^-8 per day, count = 18.242, sd = 4.49664, mean number of quakes in time period is: 340.376, Number of days in period = 10957.3

magnitude cutoff = 6.5, years = 2.03459, Probability= 4.20706*10^-7 per quake or 1.78225*10^-8 per day, count = 0.8065, sd = 0.643103, mean number of quakes in time period is: 31.481, Number of days in period = 743.117

magnitude cutoff = 6.5, years = 10., Probability= 4.0507*10^-7 per quake or 2.68901*10^-8 per day, count = 2.99625, sd = 1.50066, mean number of quakes in time period is: 121.231, Number of days in period = 3652.42

magnitude cutoff = 6.5, years = 20., Probability= 4.07773*10^-7 per quake or 2.57965*10^-8 per day, count = 5.7665, sd = 2.08564, mean number of quakes in time period is: 231.059, Number of days in period = 7304.84

magnitude cutoff = 6.5, years = 30., Probability= 4.05298*10^-7 per quake or 2.51803*10^-8 per day, count = 8.4485, sd = 2.52889, mean number of quakes in time period is: 340.376, Number of days in period = 10957.3

magnitude cutoff = 7., years = 11.976, Probability= 5.83627*10^-8 per quake or 1.92017*10^-9 per day, count = 0.6695, sd = 0.533167, mean number of quakes in time period is: 143.913, Number of days in period = 4374.16

magnitude cutoff = 7., years = 10., Probability= 5.67901*10^-8 per quake or 3.76994*10^-9 per day, count = 0.54975, sd = 0.556601, mean number of quakes in time period is: 121.231, Number of days in period = 3652.42

magnitude cutoff = 7., years = 20., Probability= 5.75417*10^-8 per quake or 3.6402*10^-9 per day, count = 1.057, sd = 0.768633, mean number of quakes in time period is: 231.059, Number of days in period = 7304.84

magnitude cutoff = 7., years = 30., Probability= 5.75641*10^-8 per quake or 3.57633*10^-9 per day, count = 1.5565, sd = 0.941807, mean number of quakes in time period is: 340.376, Number of days in period = 10957.3

magnitude cutoff = 7.5, years = 30., Probability= 0. per quake or 0. per day, count = 0., sd = 0., mean number of quakes in time period is: 342.43, Number of days in period = 10957.3

Appendix B - Calcppm - BASIC program to compute all six variables from ppm at sea level

```
Sub Calcppm()

'copyright Chondrally March 30 2009

'References, CO2 in Seawater, Equilibrium, kinetics and isotopes, by Richard E. Zeebe and Dieter Wolf-Gladrow

'also help from excel file by Ernie Lewis, Doug Wallace

'Calcppm Macro, routine to solve all six variables just from 1 variable ppm at sea surface, calibrated on 1990 IPCC data.

'Macro recorded mm=02/dd=14/yyyy=2009 by Chondrally Scientist AAAS, ASA, IEEE

'

Dim I As Integer, J As Long, Temp As Variant, LoopNum As Long, Revelle As Double
```

```
Dim c As String, HCO3 As Double, CO3 As Double, TA As Double, lastTA As Double, lastH As Double

Dim TS(3) As Double, TF(3) As Double

Dim Co As Double, Z As Double, U As Double, C0gas As Double, firstTA As Double

Dim count As Long, d As Double, e As Double, f11 As Double, f22 As Double, x11 As Double, x22 As Double, x33 As Double

Dim neighbourup As Double, flag As Boolean, fCo22nd As Double

Dim a1 As Double, a2 As Double, Result(2) As Double, a3 As Double

Dim a As Double, b As Double, C1 As Double, f As Double

Dim S1 As Double, conccoeff1 As Double, conccoeff2 As Double, ca1 As Double, ca2 As Double, astep As Double

Dim PI As Double, avtotal As Double, UF As Double, fCO2init As Double

Dim k0(3) As Double, H11 As Double

Dim N As Double, N1 As Long, Year As Double

Dim lastOmegaAr As Double, redo As Integer

Dim Av As Double, delC As Double

Dim delSolnHoverR As Double, m8 As Integer

Dim kHtheta As Double

Dim r As Double, N2 As Integer

Dim P0 As Double, firstpass1 As Integer
```

```
Dim g0 As Double

Dim gprime0 As Double

Dim T0 As Double, H31 As Double

Dim ppm0 As Double, firstTC As Double

Dim ppmasconc As Double

Dim flag2 As Integer

Dim kH As Double

Dim M As Integer

Dim delh As Double, delT As Double, num As Double, num2 As Double

Dim k1(3) As Double, k2(3) As Double, k3 As Double, k4 As Double,
kW(3) As Double, kB(3) As Double

Dim ppmschedule(200) As Double, deltaCO2(200) As Double

Dim popinit1 As Double, popinit2 As Double, popinit3 As Double, firstH
As Double

Dim OceanSat1 As Double, OceanSat2 As Double, AtmConc1 As Double,
AtmConc2 As Double, m5 As Double, m6 As Double, m7 As Double

Dim g As Double, BT(3) As Double, H1 As Double, H3 As Double, g1
As Double, g2 As Double, g3 As Double

Dim H As Double, f1 As Double, f2 As Double

Dim DIC As Double, RFgamma As Double, ds As Double, Dh As Double,
As1 As Double, Ah As Double

Dim DIC2 As Double, DIC1 As Double, S(3) As Double, DIC3 As Double
```

```
Dim pH As Double, diff As Double, m1 As Double, coeff1 As Double, coeff2 As Double

Dim m2 As Double, flag1 As Long, firstpass As Long, loop1 As Long

Dim delDICbydels2 As Double, delDICbydels1 As Double

Dim CA As Double, RPOC As Double, ksp As Double, rho(2) As Double, ksp1(3) As Double, ksp2(3) As Double

Dim calc As Double, lastca1 As Double, lastca2 As Double, dca1 As Double, dca2 As Double

Dim delca1byds As Double, delca2byds As Double, dhbyds As Double, ph1 As Double, ph2 As Double, delDICbydh1 As Double, delDICbydh2 As Double, dels As Double

Dim fCo2(3) As Double, pCO2 As Double, b1 As Double, c2 As Double, lastca11 As Double, lastca22 As Double

Dim SqrSal(3) As Double, Sal(3) As Double, TempK(3) As Double, lgTempK(3) As Double, KP1(3) As Double, KP2(3) As Double, KP3(3) As Double, KSi(3) As Double, IonS(3) As Double, lnKP1 As Double, lnKP2 As Double, lnKP3 As Double

Dim TP As Double, TotalkgSW As Double, TSi As Double

Dim KS(3) As Double, KF(3) As Double, lnKF As Double, lnKS As Double, CAlk As Double, BAlk As Double, OH As Double, PhosTop As Double, PhosBot As Double, PAlk As Double

Dim SiAlk As Double, HSO4 As Double, HF As Double, Residual As Double, Slope As Double, deltapH As Double, FREEToTOT As Double

Dim RGasConstant As Double, RT As Double, Pbar As Double, TempC As Double, kCa As Double, kAr As Double
```

```
Dim OmegaCa As Double, OmegaAr As Double, deltaVKAr As Double, KappaKAr As Double, KappaKCa As Double, deltaVKCa As Double, lnkCafac As Double, lnKArfac As Double, pHguess As Double

Dim pHTol As Double, ln10 As Double, CO32 As Double, RR As Double, Discr As Double, lastOmegaCa As Double

Dim logCO32 As Double, logkCa As Double, logkAr As Double, logcalc As Double, dig1 As Double, dig2 As Double, P(3) As Double

Dim T(3) As Double, k1a As Double, k2a As Double, kBa As Double, BTa As Double, kP1a As Double, kP2a As Double, kP3a As Double, kSia As Double, kWa As Double, TSa As Double, KSa As Double, TFa As Double

Dim KFa As Double, d1 As Double, d2 As Double, d3 As Double, fCo2a As Double, loopy As Integer, logflag As Integer, bflag As Integer, denom2 As Double

Dim LogFileName As String, FileNum As Integer, afirst As Double, specflag As Integer, av3 As Double, firstpass2 As Integer, ffac As Double, sumval As Double, sumval1 As Double, sumval2 As Double

Dim emiss1990 As Double, oceanabsorb As Double


PI = 3.14159265358979

ffac = 1#

ActiveWorkbook.ActiveSheet.Range("AF145:AF145").Cells(1) = 1#

Av = 6.0221415e+23

sigfac = 1e-06

firstpass2 = 1
```

```
logflag = 0

bflag = 0

specflag = 0


loop1 = 3

firstpass = 0

While (firstpass = 0) And (loop1 > 0)

    c = InputBox(" Enter 1 for total pass through all years, 2 for one ppm
only, 3 for one ppm surface waters only")
    a1 = c
    If (a1 < 1 Or a1 > 3) Or floor(a1) <> a1 Then
        MsgBox ("Must enter an integer 1 or 2 or 3")
        firstpass = 0
    Else
        firstpass = 1
    End If
    loop1 = loop1 - 1
Wend
    If loop1 <= 0 Then
        Return
    End If

flag2 = 1

'N1 = 37262 - 150 + 1

N1 = 3483 - 150 + 1

loopcount = 92 - 32 + 1

If a1 > 1 And a1 < 4 Then
```

```
loopcount = 1
If a1 = 2 Then
    'N1 = 37262 - 150 + 1
    N1 = 3483 - 150 + 1
Else
    N1 = 3483 - 150 + 1
End If

loop1 = 3
firstpass = 0
While (firstpass = 0) And (loop1 > 0)
    c = InputBox("Enter ppm value to use: ")
    a2 = c
If a2 < 150 Or a2 > 2000 Then
    MsgBox (" ppm must be greater than 150 and less than 2000!!")
    firstpass = 0
Else
    firstpass = 1
End If

loop1 = loop1 - 1
Wend
If loop1 <= 0 Then
    Return
End If
loop1 = 1
flag2 = 2
End If
```

```
firstpass = 1

'deltaCO2 is the amount total CO2 in the atmosphere and ocean has increased in one year

'deltaMANCO2 is the amount that has been emitted antrhopomorhically within one year.... it is less than deltaCO2
```

```
loopcount = 121 - 32 + 1
count = 1
While (count < loopcount)
If firstpass2 = 1 Then
    count = 1
    sumval = 0#
For z1 = 1 To 48 - 32 + 1
    sumval = sumval + Range("N32:N100").Cells(z1)
Next
    emiss1990 = 800# - 1# / 3# * sumval
ElseIf firstpass2 = 2 Then
    count = 48 - 32 + 1
Else
    sumval = 0
For z1 = 1 To count
    sumval = sumval + Range("N32:N100").Cells(z1)
Next
    oceanabsorb = 1# / 3# * sumval
    ffac = (emiss1990 + oceanabsorb) / (sumval1 + (count - 1) * (sumval2 -
    sumval1) / (2006 - 1990))
    Range("AF145:AF145").Cells(1) = ffac
    Range("W32:W95").Cells(count) = ffac
    ActiveWorkbook.ActiveSheet.Calculate
End If

N1 = 10150 - 150 + 1
Year = Range("L32:L150").Cells(count)
If flag2 = 2 And firstpass = 1 Then
    ppmschedule(count) = a2
Else
    ppmschedule(count) = Range("J32:J142").Cells(count)
End If

ActiveWorkbook.ActiveSheet.Range("R21:R21").Cells(1) = ppmschedule(count)
ActiveWorkbook.ActiveSheet.Calculate
For J = 1 To 200000
```

```
Next
avtotal = 1

k8 = N1

Av2 = 1
Av1 = 1

ppm = ppmschedule(count)
pH=8.851267595114-0.00196832541691472*ppm+1.0609893500682e-
06 * (ppm) ^ 2

H3 = 10 ^ (-pH)
lastH = H3
firstH = H3

DIC2 = 0

firstpass = 1
firstpass1 = 1
TotalkgSW = ActiveWorkbook.ActiveSheet.Range(“BT148:BT148”).
Cells(1)

J = 1
While J < N1 - 1
If logflag = 1 Or bflag = 1 Or specflag = 1 Then
    LogFileName = “excellogfile” & J & “.txt”
    FileNum = FreeFile ‘ next file number
    Open LogFileName For Append As #FileNum ‘ creates the file if
    it doesn’t exist
End If

‘TotalkgSW = ActiveWorkbook.ActiveSheet.Range(“BT148:BT148”).
Cells(1)
    ‘MsgBox (TotalkgSW)
‘CAL = Range(“BT145:BT145”).Cells(1)
‘POC = Range(“BT138:BT138”).Cells(1)
```

```
'DOC = Range("BT136:BT136").Cells(1)

    'MsgBox (lastH)
    'MsgBox ("0. J = " & J & ", m6 = " & m6 & ", H3 = " & H3)
'If H3 <= 0 Then
    'MsgBox (H3)
    'MsgBox (J)
'End If

'pH = -log(H3) / log(10#)

TP = 8.8102e+16 / 30.97376 / TotalkgSW
TSi = 4200000000000# / TotalkgSW
If logflag Then
    Print #FileNum, "TP = " & TP & ", TSi = " & TSi
End If
For m5 = 1 To 3
    k0(m5) = Range("BN150:BN14994").Cells(J + m5 - 1)
    k1(m5) = Range("AH150:AH14994").Cells(J + m5 - 1)
    k2(m5) = Range("AI150:AI14994").Cells(J + m5 - 1)
    kW(m5) = Range("AJ150:AJ14994").Cells(J + m5 - 1)
    S(m5) = Range("AG150:AG14994").Cells(J + m5 - 1)
    Sal(m5) = S(m5)
    BT(m5) = 0.000416 * S(m5) / 35
    T(m5) = Range("AC150:AC14994").Cells(J + m5 - 1)
    TempK(m5) = T(m5)

kB(m5) = Range("AK150:AK14994").Cells(J + m5 - 1)
ksp1(m5) = Range("AL150:AL14994").Cells(J + m5 - 1)
ksp2(m5) = Range("AM150:AM14994").Cells(J + m5 - 1)
P(m5) = Range("N150:N14994").Cells(J + m5 - 1)

fCo2(m5) = Range("BR150:BR14994").Cells(J + m5 - 1)

SqrSal(m5) = Sal(m5) ^ 0.5
If logflag = 1 Then
    Print #FileNum, "k0(" & m5 & "), = " & k0(m5)
```

```
    Print #FileNum, "k1(" & m5 & "), = " & k1(m5)
    Print #FileNum, "k2(" & m5 & "), = " & k2(m5)
    Print #FileNum, "kW(" & m5 & "), = " & kW(m5)
    Print #FileNum, "S(" & m5 & "), = " & S(m5)
    Print #FileNum, "Sal(" & m5 & "), = " & Sal(m5)
    Print #FileNum, "BT(" & m5 & "), = " & BT(m5)
    Print #FileNum, "T(" & m5 & "), = " & T(m5)
    Print #FileNum, "TempK(" & m5 & "), = " & TempK(m5)
    Print #FileNum, "kB(" & m5 & "), = " & kB(m5)
    Print #FileNum, "ksp1(" & m5 & "), = " & ksp1(m5)
    Print #FileNum, "ksp2(" & m5 & "), = " & ksp2(m5)
    Print #FileNum, "P(" & m5 & "), = " & P(m5)
    Print #FileNum, "fCo2(" & m5 & "), = " & fCo2(m5)
    Print #FileNum, "SqrSal(" & m5; ") = " & SqrSal(m5)
End If

lgTempK(m5) = log(T(m5))
TF(m5) = (6.7e-05 / 18.998) * (Sal(m5) / 1.80655): ' in mol/kg-SW
TS(m5) = (0.14 / 96.062) * (Sal(m5) / 1.80655)

IonS(m5) = 19.924 * Sal(m5) / (1000 - 1.005 * Sal(m5))

lnKP1 = -4576.752 / TempK(m5) + 115.54 - 18.453 * lgTempK(m5)
lnKP1 = lnKP1 + (-106.736 / TempK(m5) + 0.69171) * SqrSal(m5)
lnKP1 = lnKP1 + (-0.65643 / TempK(m5) - 0.01844) * Sal(m5)
KP1(m5) = Exp(lnKP1)
'
'
lnKP2 = -8814.715 / TempK(m5) + 172.1033 - 27.927 * lgTempK(m5)
lnKP2 = lnKP2 + (-160.34 / TempK(m5) + 1.3566) * SqrSal(m5)
lnKP2 = lnKP2 + (0.37335 / TempK(m5) - 0.05778) * Sal(m5)
KP2(m5) = Exp(lnKP2)
'
'
lnKP3 = -3070.75 / TempK(m5) - 18.126
lnKP3 = lnKP3 + (17.27039 / TempK(m5) + 2.81197) * SqrSal(m5)
```

```
lnKP3 = lnKP3 + (-44.99486 / TempK(m5) - 0.09984) * Sal(m5)
KP3(m5) = Exp(lnKP3)
'

'

lnKSi = -8904.2 / TempK(m5) + 117.4 - 19.334 * lgTempK(m5)
lnKSi = lnKSi + (-458.79 / TempK(m5) + 3.5913) * IonS(m5) ^ 0.5
lnKSi = lnKSi + (188.74 / TempK(m5) - 1.5998) * IonS(m5)
lnKSi = lnKSi + (-12.1652 / TempK(m5) + 0.07871) * IonS(m5) * IonS(m5)
KSi(m5) = Exp(lnKSi): ' this is on the SWS pH scale in mol/kg-H2O
KSi(m5) = KSi(m5) * (1# - 0.001005 * Sal(m5)): ' convert to mol/kg-SW
'

lnKS = -4276.1 / TempK(m5) + 141.328 - 23.093 * lgTempK(m5)
lnKS = lnKS + (-13856! / TempK(m5) + 324.57 - 47.986 * lgTempK(m5))
* IonS(m5) ^ 0.5
lnKS = lnKS + (35474! / TempK(m5) - 771.54 + 114.723 * lgTempK(m5))
* IonS(m5)
lnKS = lnKS + (-2698! / TempK(m5)) * IonS(m5) ^ 1.5
lnKS = lnKS + (1776! / TempK(m5)) * IonS(m5) * IonS(m5)
KS(m5) = Exp(lnKS): ' this is on the free pH scale in mol/kg-H2O
KS(m5) = KS(m5) * (1# - 0.001005 * Sal(m5)):
lnKF = 1590.2 / TempK(m5) - 12.641 + 1.525 * IonS(m5) ^ 0.5
KF(m5) = Exp(lnKF): ' this is on the free pH scale in mol/kg-H2O
KF(m5) = KF(m5) * (1# - 0.001005 * Sal(m5))
If logflag = 1 Then
    Print #FileNum, "TF(" & m5 & "), = " & TF(m5)
    Print #FileNum, "TS(" & m5 & "), = " & TS(m5)
    Print #FileNum, "IonS(" & m5 & "), = " & IonS(m5)
    Print #FileNum, "kp1(" & m5 & "), = " & KP1(m5)
    Print #FileNum, "kp2(" & m5 & "), = " & KP2(m5)
    Print #FileNum, "kp3(" & m5 & "), = " & KP3(m5)
    Print #FileNum, "kSi(" & m5 & "), = " & KSi(m5)
    Print #FileNum, "kS(" & m5 & "), = " & KS(m5)
    Print #FileNum, "kF(" & m5 & "), = " & KF(m5)
End If
Next
    m5 = 1
```

```
TempC = TempK(1) - 273.15
'd (A)
a = k0(1) * fCo2(1) * Av2
ainit = a
astep = a

If firstpass = 1 Then
    diffDIC = dela
    diffTA = 0
    a1 = 0
    lastH = H3
    'MsgBox ("1st lastH = " & lastH)
    a1 = 0
    x11 = 0.084
    Av2 = x11
    x22 = 1.001 * Av2
    redo = 0
    loopy = 0
    fCO2init = fCo2(1)
    loop1: If redo = 1 Then

a1 = 0

lastH = firstH
firstpass = 0
redo = 0
    'MsgBox ("Revelle = " & Revelle)
    'MsgBox ("UF = " & UF)
If logflag = 1 Then
    Print #FileNum, "Revelle1 = " & Revelle
    Print #FileNum, "Av2 = " & Av2
    Print #FileNum, "Abs(x22 / x11 - 1) = " & Abs(x22 / x11 - 1)
End If

If Abs(x22 / x11 - 1) > 1e-08 Then
    redo = 1
```

```
    firstpass = 1

If loopy = 1 Then
    Av2 = Abs(x22)
    f11 = 1# - Revelle
If logflag = 1 Then
    Print #FileNum, "x22 = " & x22
    Print #FileNum, "Av2 = " & Av2
    Print #FileNum, "f11 = " & f11
End If
Else
    f22 = 1# - Revelle
If Abs(f22 - f11) > 0 Then
    x33 = x22 - (x22 - x11) * f22 / (f22 - f11)
Else
    x33 = x22
    firstpass = 0
    redo = 0
End If

    'MsgBox (" x33 = " & x33 & ", f22 = " & f22)
    Av2 = Abs(x33)

    'MsgBox ("Av2 = " & Av2)
If logflag = 1 Then
    Print #FileNum, "x33 = " & x33
    Print #FileNum, "Av2 = " & Av2
    Print #FileNum, "f22 = " & f22
End If
'If count = 2 Then
    'MsgBox ("x33 = " & x33 & ", Av2 = " & Av2 & ", f22 = " & f22)
    'MsgBox ("Revelle = " & Revelle & ", UF = " & UF)
'End If

x22 = x33
x11 = x22
```

```
f11 = f22
End If

Else
    Av2 = Av2 * Revelle * ffac
    'MsgBox ("av2= " & Av2)
End If

ainit = k0(1) * fCo2(1) * Av2
astep = ainit
a = ainit
loopy = loopy + 1
End If

    'MsgBox ("2nd lastH = " & lastH)

HCO3 = a * k1(1) / lastH
CO3 = a * k1(1) * k2(1) / (lastH * lastH)
DIC3 = a + HCO3 + CO3
firstTC = DIC3
lastDIC = DIC3
    'Msgbox ("lastDIC = " & lastDIC)

CAlk = HCO3 + 2 * CO3
BAlk = BT(1) * kB(1) / (kB(1) + lastH)
OH = kW(1) / lastH
PhosTop = KP1(1) * KP2(1) * lastH + 2# * KP1(1) * KP2(1) * KP3(1) -
lastH * lastH * lastH
PhosBot = lastH * lastH * lastH + KP1(1) * lastH * lastH + KP1(1) *
KP2(1) * lastH + KP1(1) * KP2(1) * KP3(1)
PAlk = TP * PhosTop / PhosBot
SiAlk = TSi * KSi(1) / (KSi(1) + lastH)
FREEToTOT = (1# + TS(1) / KS(1)) ' pH scale conversion factor
Hfree = lastH / FREEToTOT: ' for H on the total scale
HSO4 = TS(1) / (1# + KS(1) / Hfree) ' since KS is on the free scale
HF = TF(1) / (1# + KF(1) / Hfree) ' since KF is on the free scale
TA = CAlk + BAlk + OH + PAlk + SiAlk - Hfree - HSO4 - HF
```

```
firstTA = TA
lastTA = TA

d = 2 * k0(1) + 2 * k0(3) - 4 * k0(2)
e = -3 * k0(1) - k0(3) + 4 * k0(2)
f = k0(1)

k0a = d * (a1) ^ 2 + e * (a1) + f
d = 2 * k1(1) + 2 * k1(3) - 4 * k1(2)
e = -3 * k1(1) - k1(3) + 4 * k1(2)
f = k1(1)

k1a = d * (a1) ^ 2 + e * (a1) + f
d = 2 * k2(1) + 2 * k2(3) - 4 * k2(2)
e = -3 * k2(1) - k2(3) + 4 * k2(2)
f = k2(1)

k2a = d * (a1) ^ 2 + e * (a1) + f

d = 2 * BT(1) + 2 * BT(3) - 4 * BT(2)
e = -3 * BT(1) - BT(3) + 4 * BT(2)
f = BT(1)

BTa = d * (a1) ^ 2 + e * (a1) + f
d = 2 * kB(1) + 2 * kB(3) - 4 * kB(2)
e = -3 * kB(1) - kB(3) + 4 * kB(2)
f = kB(1)

kBa = d * (a1) ^ 2 + e * (a1) + f

d = 2 * KP1(1) + 2 * KP1(3) - 4 * KP1(2)
e = -3 * KP1(1) - KP1(3) + 4 * KP1(2)
f = KP1(1)

kP1a = d * (a1) ^ 2 + e * (a1) + f

d = 2 * KP2(1) + 2 * KP2(3) - 4 * KP2(2)
```

```
e = -3 * KP2(1) - KP2(3) + 4 * KP2(2)
f = KP2(1)

kP2a = d * (a1) ^ 2 + e * (a1) + f

d = 2 * KP3(1) + 2 * KP3(3) - 4 * KP3(2)
e = -3 * KP3(1) - KP3(3) + 4 * KP3(2)
f = KP3(1)

kP3a = d * (a1) ^ 2 + e * (a1) + f

d = 2 * KSi(1) + 2 * KSi(3) - 4 * KSi(2)
e = -3 * KSi(1) - KSi(3) + 4 * KSi(2)
f = KSi(1)

kSia = d * (a1) ^ 2 + e * (a1) + f
d = 2 * TS(1) + 2 * TS(3) - 4 * TS(2)
e = -3 * TS(1) - TS(3) + 4 * TS(2)
f = TS(1)

TSa = d * (a1) ^ 2 + c * (a1) + f

d = 2 * KS(1) + 2 * KS(3) - 4 * KS(2)
e = -3 * KS(1) - KS(3) + 4 * KS(2)
f = KS(1)

KSa = d * (a1) ^ 2 + e * (a1) + f

d = 2 * TF(1) + 2 * TF(3) - 4 * TF(2)
e = -3 * TF(1) - TF(3) + 4 * TF(2)
f = TF(1)

TFa = d * (a1) ^ 2 + e * (a1) + f
d = 2 * KF(1) + 2 * KF(3) - 4 * KF(2)
e = -3 * KF(1) - KF(3) + 4 * KF(2)
f = KF(1)
```

```
KFa = d * (a1) ^ 2 + e * (a1) + f
d = 2 * kW(1) + 2 * kW(3) - 4 * kW(2)
e = -3 * kW(1) - kW(3) + 4 * kW(2)
f = kW(1)

kWa = d * (a1) ^ 2 + e * (a1) + f

d = 2 * fCo2(1) + 2 * fCo2(3) - 4 * fCo2(2)
e = -3 * fCo2(1) - fCo2(3) + 4 * fCo2(2)
f = fCo2(1)

fCo2a = d * (a1) ^ 2 + e * (a1) + f

If logflag = 1 Then
    Print #FileNum, "k0a = " & k0a
    Print #FileNum, "k1a = " & k1a
    Print #FileNum, "k2a = " & k2a
    Print #FileNum, "BTa = " & BTa;
    Print #FileNum, "kp1a = " & kP1a
    Print #FileNum, "kp2a = " & kP2a
    Print #FileNum, "kp3a = " & kP3a
    Print #FileNum, "kBa = " & kBa
    Print #FileNum, "kSia = " & kSia
    Print #FileNum, "TSa = " & TSa
    Print #FileNum, "kSa = " & KSa
    Print #FileNum, "TFa = " & TFa
    Print #FileNum, "kFa = " & KFa
    Print #FileNum, "kWa = " & kWa
    Print #FileNum, "fCo2a = " & fCo2a
End If

denom = (-k1a * a / lastH ^ 2 - 4 * k1a * k2a / lastH ^ 3 * a)

enom = denom - BTa * kBa / (kBa + lastH) ^ 2
a2 = lastH ^ 3 + kP1a * lastH ^ 2 + kP1a * kP2a * lastH + kP1a * kP2a * kP3a
denom = denom + TP * (kP1a * kP2a - 3 * lastH ^ 2) / a2
If logflag = 1 & bflag = 1 Then
```

```
    Print #FileNum, "count = " & count & ", J = " & J & ",m6 = " & m6
    Print #FileNum, "denom = " & denom & ", TP = " & TP & ", kp1
    = " & kP1a & ", kp2 = " & kP2a & ", kp3 = " & kP3a & ", lastH =
    " & lastH & ", a2 = " & a2
    Print #FileNum, (TP * (kP1a * kP2a * lastH + 2 * kP1a * kP2a
    * kP3a - lastH ^ 3) / a2 ^ 2 * (3 * lastH ^ 2 + 2 * kP1a * lastH +
    kP1a * kP2a))
End If
denom = denom - TSi * kSia / (kSia + lastH) ^ 2 - kWa / lastH ^ 2 - 1
/ (1 + TSa / KSa)

denom = denom - (TSa + TFa) * KFa * (1 + TSa / KSa) / (lastH + KFa
* (1 + TSa / KSa)) ^ 2
If J < 20000 Then
    denom2 = -(TP * (kP1a * kP2a * lastH + 2 * kP1a * kP2a * kP3a -
    lastH ^ 3) / a2) * ((3 * lastH ^ 2 + 2 * kP1a * lastH + kP1a * kP2a)
    / a2)
Else
    denom2 = 0
End If

If logflag = 1 & bflag = 1 Then
    Print #FileNum, " denom2 = " & denom2 & ", abs(denom2/
    denom) = " & Abs(denom2 / denom)
End If

If Abs(denom2 / denom) > 1e-06 Then
    denom = denom + denom2
End If

dela = (k0(2) * fCo2(2) - k0(1) * fCo2(1)) * Av2 / 40

x1 = 0.0001 * lastH / dela

a2 = lastH ^ 3 + kP1a * lastH ^ 2 + kP1a * kP2a * lastH + kP1a * kP2a * kP3a
f1 = -x1 * denom - (k1a / lastH + 2 * k1a * k2a / lastH ^ 2 - x1 * (k1a *
a / lastH ^ 2 + 4 * a * k1a * k2a / lastH ^ 3 + BTa * kBa / (kBa + lastH)
```

```
^ 2 - TP * (kP1a * kP2a - 3 * lastH ^ 2) / a2 + ((TP * kP1a * kP2a *
lastH + 2 * kP1a * kP2a * kP3a - lastH ^ 2) / a2) * ((3 * lastH ^ 2 +
2 * kP1a * lastH + kP1a * kP2a) / a2) + TSia * kSia / (kSia + lastH) ^
2 + kWa / lastH ^ 2 + 1 / (1 + TSa / KSa) + (TSa + TFa) * KFa * (1 +
TSa / KSa) / (1 + KFa * (lastH + TSa / KSa)) ^ 2))
f1 = f1 + 1 / a * denom / (-k1a / lastH ^ 2 - k1a * k2a / lastH ^ 3) *
(-1 - k1a / lastH - k1a * k2a / lastH ^ 2 - a * x1 * (-k1a / lastH ^ 2 - 2
* k1a * k2a / lastH ^ 3))
x2 = x1 * 1.001

f2 = -x2 * denom - (k1a / lastH + 2 * k1a * k2a / lastH ^ 2 - x2 * (k1a
* a / lastH ^ 2 + 4 * a * k1a * k2a / lastH ^ 3 + BTa * kBa / (kBa +
lastH) ^ 2 - TP * (kP1a * kP2a - 3 * lastH ^ 2) / a2 + (TP * kP1a *
kP2a * lastH + 2 * kP1a * kP2a * kP3a - lastH ^ 2) / a2 ^ 2 * (3 * lastH
^ 2 + 2 * kP1a * lastH + kP1a * kP2a) + TSia * kSia / (kSia + lastH) ^
2 + kWa / lastH ^ 2 + 1 / (1 + TSa / KSa) + (TSa + TFa) * KFa * (1 +
TSa / KSa) / (1 + KFa * (lastH + TSa / KSa)) ^ 2))
f2 = f2 + 1 / a * denom / (-k1a / lastH ^ 2 - k1a * k2a / lastH ^ 3) *
(-1 - k1a / lastH - k1a * k2a / lastH ^ 2 - a * x2 * (-k1a / lastH ^ 2 - 2
* k1a * k2a / lastH ^ 3))

While Abs(f2 - f1) > 0 And Abs(x2 / x1 - 1) > 1e-06
x3 = x2 - (x2 - x1) * f2 / (f2 - f1)
x1 = x2
x2 = x3
f1 = f2

f2 = -x2 * denom - (k1a / lastH + 2 * k1a * k2a / lastH ^ 2 - x2 * (k1a *
a / lastH ^ 2 + 4 * a * k1a * k2a / lastH ^ 3 + BTa * kBa / (kBa + lastH)
^ 2 - (TP * (kP1a * kP2a - 3 * lastH ^ 2) / a2) + ((TP * kP1a * kP2a
* lastH + 2 * kP1a * kP2a * kP3a - lastH ^ 2) / a2) * ((3 * lastH ^ 2 +
2 * kP1a * lastH + kP1a * kP2a) / a2) + TSia * kSia / (kSia + lastH) ^
2 + kWa / lastH ^ 2 + 1 / (1 + TSa / KSa) + (TSa + TFa) * KFa * (1 +
TSa / KSa) / (1 + KFa * (lastH + TSa / KSa)) ^ 2))
```

```
f2 = f2 + 1 / a * denom / (-k1a / lastH ^ 2 - k1a * k2a / lastH ^ 3) *
(-1 - k1a / lastH - k1a * k2a / lastH ^ 2 - a * x2 * (-k1a / lastH ^ 2 - 2
* k1a * k2a / lastH ^ 3))
Wend
b1 = x2
End If

If J < N1 - 1 Then
    'MsgBox ("b1 = " & b1)
    N2 = 39
    lastTA = TA
    lastDIC = DIC3
    m6 = 0
    redo = 0
    d1 = fCo2(1) * k0(1) * Av2
    d2 = fCo2(2) * k0(2) * Av2
    d3 = fCo2(3) * k0(3) * Av2
    g1 = 2 * d1 + 2 * d3 - 4 * d2
    g2 = -3 * d1 - d3 + 4 * d2
    g3 = d1
    If firstpass = 1 Then
    fCO2init = fCo2a
End If

While m6 <= N2
If firstpass = 1 Then
If m6 = 1 Then
    fCo22nd = fCo2a
End If
End If
a1 = m6 / 40

H3 = lastH

If logflag = 1 Then
    Print #FileNum, "b1 = " & b1 & ", lastH = " & lastH
```

```
End If

'MsgBox ("KP1(1) = " & KP1(1) & ", KP1(2) = " & KP1(2) & ", KPa
= " & kP1a)

a1 = m6 / 40 / 2 + 1 / 4 / 40

a = g1 * (a1) ^ 2 + g2 * (a1) + g3

d = 2 * k0(1) + 2 * k0(3) - 4 * k0(2)
e = -3 * k0(1) - k0(3) + 4 * k0(2)
f = k0(1)

k0a = d * (a1) ^ 2 + e * (a1) + f
d = 2 * k1(1) + 2 * k1(3) - 4 * k1(2)
e = -3 * k1(1) - k1(3) + 4 * k1(2)
f = k1(1)

k1a = d * (a1) ^ 2 + e * (a1) + f
d = 2 * k2(1) + 2 * k2(3) - 4 * k2(2)
e = -3 * k2(1) - k2(3) + 4 * k2(2)
f = k2(1)

k2a = d * (a1) ^ 2 + e * (a1) + f

d = 2 * BT(1) + 2 * BT(3) - 4 * BT(2)
e = -3 * BT(1) - BT(3) + 4 * BT(2)
f = BT(1)

BTa = d * (a1) ^ 2 + e * (a1) + f
d = 2 * kB(1) + 2 * kB(3) - 4 * kB(2)
e = -3 * kB(1) - kB(3) + 4 * kB(2)
f = kB(1)

kBa = d * (a1) ^ 2 + e * (a1) + f

d = 2 * KP1(1) + 2 * KP1(3) - 4 * KP1(2)
```

```
e = -3 * KP1(1) - KP1(3) + 4 * KP1(2)
f = KP1(1)

kP1a = d * (a1) ^ 2 + e * (a1) + f

d = 2 * KP2(1) + 2 * KP2(3) - 4 * KP2(2)
e = -3 * KP2(1) - KP2(3) + 4 * KP2(2)
f = KP2(1)

kP2a = d * (a1) ^ 2 + e * (a1) + f

d = 2 * KP3(1) + 2 * KP3(3) - 4 * KP3(2)
e = -3 * KP3(1) - KP3(3) + 4 * KP3(2)
f = KP3(1)

kP3a = d * (a1) ^ 2 + e * (a1) + f

d = 2 * KSi(1) + 2 * KSi(3) - 4 * KSi(2)
e = -3 * KSi(1) - KSi(3) + 4 * KSi(2)
f = KSi(1)

kSia = d * (a1) ^ 2 + e * (a1) + f
d = 2 * TS(1) + 2 * TS(3) - 4 * TS(2)
e = -3 * TS(1) - TS(3) + 4 * TS(2)
f = TS(1)

TSa = d * (a1) ^ 2 + e * (a1) + f

d = 2 * KS(1) + 2 * KS(3) - 4 * KS(2)
e = -3 * KS(1) - KS(3) + 4 * KS(2)
f = KS(1)

KSa = d * (a1) ^ 2 + e * (a1) + f

d = 2 * TF(1) + 2 * TF(3) - 4 * TF(2)
e = -3 * TF(1) - TF(3) + 4 * TF(2)
f = TF(1)
```

```
TFa = d * (a1) ^ 2 + e * (a1) + f
d = 2 * KF(1) + 2 * KF(3) - 4 * KF(2)
e = -3 * KF(1) - KF(3) + 4 * KF(2)
f = KF(1)

KFa = d * (a1) ^ 2 + e * (a1) + f
d = 2 * kW(1) + 2 * kW(3) - 4 * kW(2)
e = -3 * kW(1) - kW(3) + 4 * kW(2)
f = kW(1)

kWa = d * (a1) ^ 2 + e * (a1) + f

d = 2 * fCo2(1) + 2 * fCo2(3) - 4 * fCo2(2)
e = -3 * fCo2(1) - fCo2(3) + 4 * fCo2(2)
f = fCo2(1)

fCo2a = d * (a1) ^ 2 + e * (a1) + f

If logflag = 1 Then
    Print #FileNum, " a = " & a
    Print #FileNum, "k0a = " & k0a
    Print #FileNum, "k1a = " & k1a
    Print #FileNum, "k2a = " & k2a
    Print #FileNum, "BTa = " & BTa;
    Print #FileNum, "kp1a = " & kP1a
    Print #FileNum, "kp2a = " & kP2a
    Print #FileNum, "kp3a = " & kP3a
    Print #FileNum, "kBa = " & kBa
    Print #FileNum, "kSia = " & kSia
    Print #FileNum, "TSa = " & TSa
    Print #FileNum, "kSa = " & KSa
    Print #FileNum, "TFa = " & TFa
    Print #FileNum, "kFa = " & KFa
    Print #FileNum, "kWa = " & kWa
    Print #FileNum, "fCo2a = " & fCo2a
End If
```

```
'MsgBox ("a = " & a)
'MsgBox ("ainit = " & ainit)
'MsgBox ("b1 = " & b1)
'MsgBox (" H3 = " & H3)

delh = Abs(a - astep) * b1
lastH = H3 + delh

If logflag = 1 & bflag = 1 Then
    Print #FileNum, "2. b1 = " & b1 & ", lastH = " & lastH
End If

x1 = b1
a = g1 * (a1) ^ 2 + g2 * (a1) + g3

'MsgBox (lastH)

'MsgBox ("3rd lastH = " & lastH)
DIC2 = DIC3
HCO3 = a * k1a / lastH
CO3 = k1a * k2a * a / (lastH * lastH)
DIC3 = a + HCO3 + CO3

CAlk = HCO3 + 2 * CO3
BAlk = BTa * kBa / (kBa + lastH)
OH = kWa / lastH
PhosTop = kP1a * kP2a * lastH + 2# * kP1a * kP2a * kP3a - lastH *
lastH * lastH
PhosBot = lastH * lastH * lastH + kP1a * lastH * lastH + kP1a * kP2a
* lastH + kP1a * kP2a * kP3a
PAlk = TP * PhosTop / PhosBot
SiAlk = TSi * kSia / (kSia + lastH)
FREEToTOT = (1 + TSa / KSa) ' pH scale conversion factor
Hfree = lastH / FREEToTOT ' for H on the total scale
HSO4 = TSa / (1 + KSa / Hfree) ' since KS is on the free scale
HF = TFa / (1 + KFa / Hfree) ' since KF is on the free scale
TA = CAlk + BAlk + OH + PAlk + SiAlk - Hfree - HSO4 - HF
```

```
If specflag = 1 Then
    Print #FileNum, "count = " & count & ", J = " & J & ",m6 = " & m6
    Print #FileNum, ",k1a = " & k1a & ", k2a = " & k2a & ",lastH =
    " & lastH & ", a = " & a
    Print #FileNum, (-k1a * a / lastH ^ 2 - 4 * k1a * k2a / lastH ^ 3 * a)
End If
denom = (-k1a * a / lastH ^ 2 - 4 * k1a * k2a / lastH ^ 3 * a)

denom = denom - BTa * kBa / (kBa + lastH) ^ 2
a2 = lastH ^ 3 + kP1a * lastH ^ 2 + kP1a * kP2a * lastH + kP1a *
kP2a * kP3a
denom = denom + TP * (kP1a * kP2a - 3 * lastH ^ 2) / a2
If logflag = 1 & bflag = 1 Then
    Print #FileNum, "count = " & count & ", J = " & J & ",m6 = " & m6
    Print #FileNum, "denom = " & denom & ", TP = " & TP & ",
    kp1 = " & kP1a & ", kp2 = " & kP2a & ", kp3 = " & kP3a & ",
    lastH = " & lastH & ", a2 = " & a2
    Print #FileNum, (TP * (kP1a * kP2a * lastH + 2 * kP1a * kP2a
    * kP3a - lastH ^ 3) / a2 ^ 2 * (3 * lastH ^ 2 + 2 * kP1a * lastH +
    kP1a * kP2a))
End If
denom = denom - TSi * kSia / (kSia + lastH) ^ 2 - kWa / lastH ^ 2 - 1
/ (1 + TSa / KSa)

denom = denom - (TSa + TFa) * KFa * (1 + TSa / KSa) / (lastH + KFa
* (1 + TSa / KSa)) ^ 2
If J < 20000 Then
    denom2 = -(TP * (kP1a * kP2a * lastH + 2 * kP1a * kP2a * kP3a -
    lastH ^ 3) / a2) * ((3 * lastH ^ 2 + 2 * kP1a * lastH + kP1a * kP2a) / a2)
Else
denom2 = 0
End If
If logflag = 1 & bflag = 1 Then
    Print #FileNum, " denom2 = " & denom2 & ", abs(denom2/
    denom) = " & Abs(denom2 / denom)
End If
```

```
If Abs(denom2 / denom) > 1e-06 Then
    denom = denom + denom2
End If

dela = Abs(a - astep)
a2 = lastH ^ 3 + kP1a * lastH ^ 2 + kP1a * kP2a * lastH + kP1a * kP2a * kP3a
f1 = -x1 * denom - (-(TA - lastTA) / dela + k1a / lastH + 2 * k1a * k2a
/ lastH ^ 2 - x1 * (k1a * a / lastH ^ 2 + 4 * a * k1a * k2a / lastH ^ 3 +
BTa * kBa / (kBa + lastH) ^ 2 - TP * (kP1a * kP2a - 3 * lastH ^ 2) /
a2 + ((TP * kP1a * kP2a * lastH + 2 * kP1a * kP2a * kP3a - lastH ^ 2)
/ a2) * ((3 * lastH ^ 2 + 2 * kP1a * lastH + kP1a * kP2a) / a2) + TSia *
kSia / (kSia + lastH) ^ 2 + kWa / lastH ^ 2 + 1 / (1 + TSa / KSa) + (TSa
+ TFa) * KFa * (1 + TSa / KSa) / (1 + KFa * (lastH + TSa / KSa)) ^ 2))
f1 = f1 + 1 / a * denom / (-k1a / lastH ^ 2 - k1a * k2a / lastH ^ 3) *
((DIC3 - lastDIC) / dela - 1 - k1a / lastH - k1a * k2a / lastH ^ 2 - a *
x1 * (-k1a / lastH ^ 2 - 2 * k1a * k2a / lastH ^ 3))
x2 = x1 * 1.001

f2 = -x2 * denom - (-(TA - lastTA) / dela + k1a / lastH + 2 * k1a * k2a
/ lastH ^ 2 - x2 * (k1a * a / lastH ^ 2 + 4 * a * k1a * k2a / lastH ^ 3 +
BTa * kBa / (kBa + lastH) ^ 2 - TP * (kP1a * kP2a - 3 * lastH ^ 2) /
a2 + (TP * kP1a * kP2a * lastH + 2 * kP1a * kP2a * kP3a - lastH ^ 2) /
a2 ^ 2 * (3 * lastH ^ 2 + 2 * kP1a * lastH + kP1a * kP2a) + TSia * kSia
/ (kSia + lastH) ^ 2 + kWa / lastH ^ 2 + 1 / (1 + TSa / KSa) + (TSa +
TFa) * KFa * (1 + TSa / KSa) / (1 + KFa * (lastH + TSa / KSa)) ^ 2))
f2 = f2 + 1 / a * denom / (-k1a / lastH ^ 2 - k1a * k2a / lastH ^ 3) *
((DIC3 - lastDIC) / dela - 1 - k1a / lastH - k1a * k2a / lastH ^ 2 - a *
x2 * (-k1a / lastH ^ 2 - 2 * k1a * k2a / lastH ^ 3))

While Abs(f2 - f1) > 0 And Abs(x2 / x1 - 1) > 1e-06
x3 = x2 - (x2 - x1) * f2 / (f2 - f1)
x1 = x2
x2 = x3
f1 = f2
```

```
f2 = -x2 * denom - (-(TA - lastTA) / dela + k1a / lastH + 2 * k1a * k2a
/ lastH ^ 2 - x2 * (k1a * a / lastH ^ 2 + 4 * a * k1a * k2a / lastH ^ 3 +
BTa * kBa / (kBa + lastH) ^ 2 - (TP * (kP1a * kP2a - 3 * lastH ^ 2) /
a2) + ((TP * kP1a * kP2a * lastH + 2 * kP1a * kP2a * kP3a - lastH ^ 2)
/ a2) * ((3 * lastH ^ 2 + 2 * kP1a * lastH + kP1a * kP2a) / a2) + TSia *
kSia / (kSia + lastH) ^ 2 + kWa / lastH ^ 2 + 1 / (1 + TSa / KSa) + (TSa
+ TFa) * KFa * (1 + TSa / KSa) / (1 + KFa * (lastH + TSa / KSa)) ^ 2))
f2 = f2 + 1 / a * denom / (-k1a / lastH ^ 2 - k1a * k2a / lastH ^ 3) *
((DIC3 - lastDIC) / dela - 1 - k1a / lastH - k1a * k2a / lastH ^ 2 - a *
x2 * (-k1a / lastH ^ 2 - 2 * k1a * k2a / lastH ^ 3))
Wend

b2 = x2
'MsgBox ("b2 = " & b2)

If logflag = 1 Then
    Print #FileNum, " b2 = " & b2
    Print #FileNum, " a = " & a
End If

delh = Abs(a - astep) * b2
lastH = H3 + delh
If logflag = 1 & bflag = 1 Then
    Print #FileNum, "2. b2 = " & b2 & ", lastH = " & lastH
End If

x1 = b2

'MsgBox ("b2 lastH = " & lastH)
'MsgBox ("4th lastH = " & lastH)
DIC2 = DIC3
HCO3 = a * k1a / lastH
CO3 = k1a * k2a * a / (lastH * lastH)
DIC3 = a + HCO3 + CO3

CAlk = HCO3 + 2 * CO3
BAlk = BTa * kBa / (kBa + lastH)
```

```
OH = kWa / lastH
PhosTop = kP1a * kP2a * lastH + 2# * kP1a * kP2a * kP3a - lastH *
lastH * lastH
PhosBot = lastH * lastH * lastH + kP1a * lastH * lastH + kP1a * kP2a
* lastH + kP1a * kP2a * kP3a
PAlk = TP * PhosTop / PhosBot
SiAlk = TSi * kSia / (kSia + lastH)
FREEToTOT = (1 + TSa / KSa) ‘ pH scale conversion factor
Hfree = lastH / FREEToTOT ‘ for H on the total scale
HSO4 = TSa / (1 + KSa / Hfree) ‘ since KS is on the free scale
HF = TFa / (1 + KFa / Hfree) ‘ since KF is on the free scale
TA = CAlk + BAlk + OH + PAlk + SiAlk - Hfree - HSO4 - HF

‘MsgBox (lastH)

denom = (-k1a * a / lastH ^ 2 - 4 * k1a * k2a / lastH ^ 3 * a)

denom = denom - BTa * kBa / (kBa + lastH) ^ 2
a2 = lastH ^ 3 + kP1a * lastH ^ 2 + kP1a * kP2a * lastH + kP1a * kP2a * kP3a
denom = denom + TP * (kP1a * kP2a - 3 * lastH ^ 2) / a2
If logflag = 1 & bflag = 1 Then
    Print #FileNum, “count = “ & count & “, J = “ & J & “,m6 = “ & m6
    Print #FileNum, “denom = “ & denom & “, TP = “ & TP & “,
    kp1 = “ & kP1a & “, kp2 = “ & kP2a & “, kp3 = “ & kP3a & “,
    lastH = “ & lastH & “, a2 = “ & a2
    Print #FileNum, (TP * (kP1a * kP2a * lastH + 2 * kP1a * kP2a
    * kP3a - lastH ^ 3) / a2 ^ 2 * (3 * lastH ^ 2 + 2 * kP1a * lastH +
    kP1a * kP2a))
End If
denom = denom - TSi * kSia / (kSia + lastH) ^ 2 - kWa / lastH ^ 2 - 1
/ (1 + TSa / KSa)

denom = denom - (TSa + TFa) * KFa * (1 + TSa / KSa) / (lastH + KFa
* (1 + TSa / KSa)) ^ 2
If J < 20000 Then
```

```
        denom2 = -(TP * (kP1a * kP2a * lastH + 2 * kP1a * kP2a * kP3a -
        lastH ^ 3) / a2) * ((3 * lastH ^ 2 + 2 * kP1a * lastH + kP1a * kP2a) / a2)
    Else
        denom2 = 0
    End If
    If logflag = 1 & bflag = 1 Then
        Print #FileNum, " denom2 = " & denom2 & ", abs(denom2/
        denom) = " & Abs(denom2 / denom)
    End If
    If Abs(denom2 / denom) > 1e-06 Then
        denom = denom + denom2
    End If

    dela = Abs(a - astep)
    a2 = lastH ^ 3 + kP1a * lastH ^ 2 + kP1a * kP2a * lastH + kP1a *
    kP2a * kP3a
    f1 = -x1 * denom - (-(TA - lastTA) / dela + k1a / lastH + 2 * k1a * k2a
    / lastH ^ 2 - x1 * (k1a * a / lastH ^ 2 + 4 * a * k1a * k2a / lastH ^ 3 +
    BTa * kBa / (kBa + lastH) ^ 2 - TP * (kP1a * kP2a - 3 * lastH ^ 2) /
    a2 + ((TP * kP1a * kP2a * lastH + 2 * kP1a * kP2a * kP3a - lastH ^ 2)
    / a2) * ((3 * lastH ^ 2 + 2 * kP1a * lastH + kP1a * kP2a) / a2) + TSia *
    kSia / (kSia + lastH) ^ 2 + kWa / lastH ^ 2 + 1 / (1 + TSa / KSa) + (TSa
    + TFa) * KFa * (1 + TSa / KSa) / (1 + KFa * (lastH + TSa / KSa)) ^ 2))
    f1 = f1 + 1 / a * denom / (-k1a / lastH ^ 2 - k1a * k2a / lastH ^ 3) *
    ((DIC3 - lastDIC) / dela - 1 - k1a / lastH - k1a * k2a / lastH ^ 2 - a *
    x1 * (-k1a / lastH ^ 2 - 2 * k1a * k2a / lastH ^ 3))
    x2 = x1 * 1.001

    f2 = -x2 * denom - (-(TA - lastTA) / dela + k1a / lastH + 2 * k1a * k2a
    / lastH ^ 2 - x2 * (k1a * a / lastH ^ 2 + 4 * a * k1a * k2a / lastH ^ 3 +
    BTa * kBa / (kBa + lastH) ^ 2 - TP * (kP1a * kP2a - 3 * lastH ^ 2) /
    a2 + (TP * kP1a * kP2a * lastH + 2 * kP1a * kP2a * kP3a - lastH ^ 2) /
    a2 ^ 2 * (3 * lastH ^ 2 + 2 * kP1a * lastH + kP1a * kP2a) + TSia * kSia
    / (kSia + lastH) ^ 2 + kWa / lastH ^ 2 + 1 / (1 + TSa / KSa) + (TSa +
    TFa) * KFa * (1 + TSa / KSa) / (1 + KFa * (lastH + TSa / KSa)) ^ 2))
```

```
f2 = f2 + 1 / a * denom / (-k1a / lastH ^ 2 - k1a * k2a / lastH ^ 3) *
((DIC3 - lastDIC) / dela - 1 - k1a / lastH - k1a * k2a / lastH ^ 2 - a *
x2 * (-k1a / lastH ^ 2 - 2 * k1a * k2a / lastH ^ 3))

While Abs(f2 - f1) > 0 And Abs(x2 / x1 - 1) > 1e-06
x3 = x2 - (x2 - x1) * f2 / (f2 - f1)
x1 = x2
x2 = x3
f1 = f2

f2 = -x2 * denom - (-(TA - lastTA) / dela + k1a / lastH + 2 * k1a * k2a
/ lastH ^ 2 - x2 * (k1a * a / lastH ^ 2 + 4 * a * k1a * k2a / lastH ^ 3 +
BTa * kBa / (kBa + lastH) ^ 2 - (TP * (kP1a * kP2a - 3 * lastH ^ 2) /
a2) + ((TP * kP1a * kP2a * lastH + 2 * kP1a * kP2a * kP3a - lastH ^ 2)
/ a2) * ((3 * lastH ^ 2 + 2 * kP1a * lastH + kP1a * kP2a) / a2) + TSia *
kSia / (kSia + lastH) ^ 2 + kWa / lastH ^ 2 + 1 / (1 + TSa / KSa) + (TSa
+ TFa) * KFa * (1 + TSa / KSa) / (1 + KFa * (lastH + TSa / KSa)) ^ 2))
f2 = f2 + 1 / a * denom / (-k1a / lastH ^ 2 - k1a * k2a / lastH ^ 3) *
((DIC3 - lastDIC) / dela - 1 - k1a / lastH - k1a * k2a / lastH ^ 2 - a *
x2 * (-k1a / lastH ^ 2 - 2 * k1a * k2a / lastH ^ 3))
Wend
b3 = x2
'MsgBox (" b3 = " & b3)
a1 = m6 / 40 / 2 + 1 / 2 / 40

delh = Abs(a - astep) * b3
lastH = H3 + delh

If logflag = 1 & bflag = 1 Then
    Print #FileNum, "3. b3 = " & b3 & ", lastH = " & lastH
End If
a = g1 * (a1) ^ 2 + g2 * (a1) + g3

'MsgBox (a & " ainit = " & ainit)
d = 2 * k0(1) + 2 * k0(3) - 4 * k0(2)
e = -3 * k0(1) - k0(3) + 4 * k0(2)
f = k0(1)
```

```
k0a = d * (a1) ^ 2 + e * (a1) + f
d = 2 * k1(1) + 2 * k1(3) - 4 * k1(2)
e = -3 * k1(1) - k1(3) + 4 * k1(2)
f = k1(1)

k1a = d * (a1) ^ 2 + e * (a1) + f
d = 2 * k2(1) + 2 * k2(3) - 4 * k2(2)
e = -3 * k2(1) - k2(3) + 4 * k2(2)
f = k2(1)

k2a = d * (a1) ^ 2 + e * (a1) + f

d = 2 * BT(1) + 2 * BT(3) - 4 * BT(2)
e = -3 * BT(1) - BT(3) + 4 * BT(2)
f = BT(1)

BTa = d * (a1) ^ 2 + e * (a1) + f
d = 2 * kB(1) + 2 * kB(3) - 4 * kB(2)
e = -3 * kB(1) - kB(3) + 4 * kB(2)
f = kB(1)

kBa = d * (a1) ^ 2 + e * (a1) + f

d = 2 * KP1(1) + 2 * KP1(3) - 4 * KP1(2)
e = -3 * KP1(1) - KP1(3) + 4 * KP1(2)
f = KP1(1)

kP1a = d * (a1) ^ 2 + e * (a1) + f

d = 2 * KP2(1) + 2 * KP2(3) - 4 * KP2(2)
e = -3 * KP2(1) - KP2(3) + 4 * KP2(2)
f = KP2(1)

kP2a = d * (a1) ^ 2 + e * (a1) + f

d = 2 * KP3(1) + 2 * KP3(3) - 4 * KP3(2)
e = -3 * KP3(1) - KP3(3) + 4 * KP3(2)
```

```
f = KP3(1)

kP3a = d * (a1) ^ 2 + e * (a1) + f

d = 2 * KSi(1) + 2 * KSi(3) - 4 * KSi(2)
e = -3 * KSi(1) - KSi(3) + 4 * KSi(2)
f = KSi(1)

kSia = d * (a1) ^ 2 + e * (a1) + f
d = 2 * TS(1) + 2 * TS(3) - 4 * TS(2)
e = -3 * TS(1) - TS(3) + 4 * TS(2)
f = TS(1)

TSa = d * (a1) ^ 2 + e * (a1) + f

d = 2 * KS(1) + 2 * KS(3) - 4 * KS(2)
e = -3 * KS(1) - KS(3) + 4 * KS(2)
f = KS(1)

KSa = d * (a1) ^ 2 + e * (a1) + f

d = 2 * TF(1) + 2 * TF(3) - 4 * TF(2)
e = -3 * TF(1) - TF(3) + 4 * TF(2)
f = TF(1)

TFa = d * (a1) ^ 2 + e * (a1) + f
d = 2 * KF(1) + 2 * KF(3) - 4 * KF(2)
e = -3 * KF(1) - KF(3) + 4 * KF(2)
f = KF(1)

KFa = d * (a1) ^ 2 + e * (a1) + f
d = 2 * kW(1) + 2 * kW(3) - 4 * kW(2)
e = -3 * kW(1) - kW(3) + 4 * kW(2)
f = kW(1)

kWa = d * (a1) ^ 2 + e * (a1) + f
```

```
d = 2 * fCo2(1) + 2 * fCo2(3) - 4 * fCo2(2)
e = -3 * fCo2(1) - fCo2(3) + 4 * fCo2(2)
f = fCo2(1)

fCo2a = d * (a1) ^ 2 + e * (a1) + f
If logflag = 1 Then
    Print #FileNum, "k0a = " & k0a
    Print #FileNum, "k1a = " & k1a
    Print #FileNum, "k2a = " & k2a
    Print #FileNum, "BTa = " & BTa;
    Print #FileNum, "kp1a = " & kP1a
    Print #FileNum, "kp2a = " & kP2a
    Print #FileNum, "kp3a = " & kP3a
    Print #FileNum, "kBa = " & kBa
    Print #FileNum, "kSia = " & kSia
    Print #FileNum, "TSa = " & TSa
    Print #FileNum, "kSa = " & KSa
    Print #FileNum, "TFa = " & TFa
    Print #FileNum, "kFa = " & KFa
    Print #FileNum, "kWa = " & kWa
    Print #FileNum, "fCo2a = " & fCo2a
End If

x1 = b3
'MsgBox ("5th lastH = " & lastH)
DIC2 = DIC3
HCO3 = a * k1a / lastH
CO3 = k1a * k2a * a / (lastH * lastH)
DIC3 = a + HCO3 + CO3

CAlk = HCO3 + 2 * CO3
BAlk = BTa * kBa / (kBa + lastH)
OH = kWa / lastH
PhosTop = kP1a * kP2a * lastH + 2# * kP1a * kP2a * kP3a - lastH *
lastH * lastH
```

```
PhosBot = lastH * lastH * lastH + kP1a * lastH * lastH + kP1a * kP2a
* lastH + kP1a * kP2a * kP3a
PAlk = TP * PhosTop / PhosBot
SiAlk = TSi * kSia / (kSia + lastH)
FREEToTOT = (1 + TSa / KSa) ‘ pH scale conversion factor
Hfree = lastH / FREEToTOT ‘ for H on the total scale
HSO4 = TSa / (1 + KSa / Hfree) ‘ since KS is on the free scale
HF = TFa / (1 + KFa / Hfree) ‘ since KF is on the free scale
TA = CAlk + BAlk + OH + PAlk + SiAlk - Hfree - HSO4 - HF

denom = (-k1a * a / lastH ^ 2 - 4 * k1a * k2a / lastH ^ 3 * a)

denom = denom - BTa * kBa / (kBa + lastH) ^ 2
a2 = lastH ^ 3 + kP1a * lastH ^ 2 + kP1a * kP2a * lastH + kP1a *
kP2a * kP3a
denom = denom + TP * (kP1a * kP2a - 3 * lastH ^ 2) / a2
If logflag = 1 Then
    Print #FileNum, “count = “ & count & “, J = “ & J & “,m6 = “ & m6
    Print #FileNum, “denom = “ & denom & “, TP = “ & TP & “,
    kp1 = “ & kP1a & “, kp2 = “ & kP2a & “, kp3 = “ & kP3a & “,
    lastH = “ & lastH & “, a2 = “ & a2
    Print #FileNum, (TP * (kP1a * kP2a * lastH + 2 * kP1a * kP2a
    * kP3a - lastH ^ 3) / a2 ^ 2 * (3 * lastH ^ 2 + 2 * kP1a * lastH +
    kP1a * kP2a))
End If
denom = denom - TSi * kSia / (kSia + lastH) ^ 2 - kWa / lastH ^ 2 - 1
/ (1 + TSa / KSa)

denom = denom - (TSa + TFa) * KFa * (1 + TSa / KSa) / (lastH + KFa
* (1 + TSa / KSa)) ^ 2
If J < 20000 Then
    denom2 = -(TP * (kP1a * kP2a * lastH + 2 * kP1a * kP2a * kP3a -
    lastH ^ 3) / a2) * ((3 * lastH ^ 2 + 2 * kP1a * lastH + kP1a * kP2a) / a2)
Else
    denom2 = 0
End If
```

```
If logflag = 1 & bflag = 1 Then
    Print #FileNum, " denom2 = " & denom2 & ", abs(denom2/
    denom) = " & Abs(denom2 / denom)
End If
If Abs(denom2 / denom) > 1e-06 Then
    denom = denom + denom2
End If

dela = Abs(a - astep)
a2 = lastH ^ 3 + kP1a * lastH ^ 2 + kP1a * kP2a * lastH + kP1a *
kP2a * kP3a
f1 = -x1 * denom - (-(TA - lastTA) / dela + k1a / lastH + 2 * k1a * k2a
/ lastH ^ 2 - x1 * (k1a * a / lastH ^ 2 + 4 * a * k1a * k2a / lastH ^ 3 +
BTa * kBa / (kBa + lastH) ^ 2 - TP * (kP1a * kP2a - 3 * lastH ^ 2) /
a2 + ((TP * kP1a * kP2a * lastH + 2 * kP1a * kP2a * kP3a - lastH ^ 2)
/ a2) * ((3 * lastH ^ 2 + 2 * kP1a * lastH + kP1a * kP2a) / a2) + TSia *
kSia / (kSia + lastH) ^ 2 + kWa / lastH ^ 2 + 1 / (1 + TSa / KSa) + (TSa
+ TFa) * KFa * (1 + TSa / KSa) / (1 + KFa * (lastH + TSa / KSa)) ^ 2))
f1 = f1 + 1 / a * denom / (-k1a / lastH ^ 2 - k1a * k2a / lastH ^ 3) *
((DIC3 - lastDIC) / dela - 1 - k1a / lastH - k1a * k2a / lastH ^ 2 - a *
x1 * (-k1a / lastH ^ 2 - 2 * k1a * k2a / lastH ^ 3))
x2 = x1 * 1.001

f2 = -x2 * denom - (-(TA - lastTA) / dela + k1a / lastH + 2 * k1a * k2a
/ lastH ^ 2 - x2 * (k1a * a / lastH ^ 2 + 4 * a * k1a * k2a / lastH ^ 3 +
BTa * kBa / (kBa + lastH) ^ 2 - TP * (kP1a * kP2a - 3 * lastH ^ 2) /
a2 + (TP * kP1a * kP2a * lastH + 2 * kP1a * kP2a * kP3a - lastH ^ 2) /
a2 ^ 2 * (3 * lastH ^ 2 + 2 * kP1a * lastH + kP1a * kP2a) + TSia * kSia
/ (kSia + lastH) ^ 2 + kWa / lastH ^ 2 + 1 / (1 + TSa / KSa) + (TSa +
TFa) * KFa * (1 + TSa / KSa) / (1 + KFa * (lastH + TSa / KSa)) ^ 2))
f2 = f2 + 1 / a * denom / (-k1a / lastH ^ 2 - k1a * k2a / lastH ^ 3) *
((DIC3 - lastDIC) / dela - 1 - k1a / lastH - k1a * k2a / lastH ^ 2 - a *
x2 * (-k1a / lastH ^ 2 - 2 * k1a * k2a / lastH ^ 3))

While Abs(f2 - f1) > 0 And Abs(x2 / x1 - 1) > 1e-06
x3 = x2 - (x2 - x1) * f2 / (f2 - f1)
```

```
x1 = x2
x2 = x3
f1 = f2

f2 = -x2 * denom - (-(TA - lastTA) / dela + k1a / lastH + 2 * k1a * k2a
/ lastH ^ 2 - x2 * (k1a * a / lastH ^ 2 + 4 * a * k1a * k2a / lastH ^ 3 +
BTa * kBa / (kBa + lastH) ^ 2 - (TP * (kP1a * kP2a - 3 * lastH ^ 2) /
a2) + ((TP * kP1a * kP2a * lastH + 2 * kP1a * kP2a * kP3a - lastH ^ 2)
/ a2) * ((3 * lastH ^ 2 + 2 * kP1a * lastH + kP1a * kP2a) / a2) + TSia *
kSia / (kSia + lastH) ^ 2 + kWa / lastH ^ 2 + 1 / (1 + TSa / KSa) + (TSa
+ TFa) * KFa * (1 + TSa / KSa) / (1 + KFa * (lastH + TSa / KSa)) ^ 2))
f2 = f2 + 1 / a * denom / (-k1a / lastH ^ 2 - k1a * k2a / lastH ^ 3) *
((DIC3 - lastDIC) / dela - 1 - k1a / lastH - k1a * k2a / lastH ^ 2 - a *
x2 * (-k1a / lastH ^ 2 - 2 * k1a * k2a / lastH ^ 3))
Wend
b4 = x2
'MsgBox ("b4 = " & b4)

If logflag = 1 Then
    Print #FileNum, " b4 = " & b4
    Print #FileNum, " a = " & a
End If

delh = Abs(a - astep) / 6 * (b1 + 2 * b2 + 2 * b3 + b4)
lastH = H3 + delh
If logflag = 1 & bflag = 1 Then
    Print #FileNum, "4. b4 = " & b4 & ", lastH = " & lastH
End If

'MsgBox ("1. J = " & J & ", m6 = " & m6 & ", lastH = " & lastH)
'MsgBox ("6th and one loop lastH = " & lastH)
'MsgBox ("one loop b = " & 1 / 6 * (b1 + 2 * b2 + 2 * b3 + b4))
'MsgBox (" a = " & a)
'MsgBox (" ainit = " & ainit)
'MsgBox ("pH3 = " & -log(H3) / log(10#) & ", plastH = " & -log(lastH)
/ log(10#))
```

```
DIC2 = DIC3
HCO3 = a * k1a / lastH
CO3 = k1a * k2a * a / (lastH * lastH)
DIC3 = a + HCO3 + CO3
diffDIC = DIC3 - lastDIC
'MsgBox ("DIC3 = " & DIC3)

CAlk = HCO3 + 2 * CO3
BAlk = BTa * kBa / (kBa + lastH)
OH = kWa / lastH
PhosTop = kP1a * kP2a * lastH + 2# * kP1a * kP2a * kP3a - lastH *
lastH * lastH
PhosBot = lastH * lastH * lastH + kP1a * lastH * lastH + kP1a * kP2a
* lastH + kP1a * kP2a * kP3a
PAlk = TP * PhosTop / PhosBot
SiAlk = TSi * kSia / (kSia + lastH)
FREEToTOT = (1 + TSa / KSa) ' pH scale conversion factor
Hfree = lastH / FREEToTOT ' for H on the total scale
HSO4 = TSa / (1 + KSa / Hfree) ' since KS is on the free scale
HF = TFa / (1 + KFa / Hfree) ' since KF is on the free scale
TA = CAlk + BAlk + OH + PAlk + SiAlk - Hfree - HSO4 - HF

diffTA = TA - lastTA

b1 = b4

'MsgBox (DIC3 & ", lastDIC = " & lastDIC)
'MsgBox (k0(2) * fCo2(2) & " a = " & a)
If firstpass = 1 Then
    Revelle = ((a - ainit) / ainit / (DIC3 - lastDIC) * lastDIC)
    UF = (a - ainit) / (fCo22nd - fCO2init)
    'MsgBox ("Revelle = " & Revelle)
    'MsgBox ("UF = " & UF)
    'MsgBox (Av2)
    redo = 1
GoTo loop1
```

```
    End If
    lastTA = TA
    lastDIC = DIC3
    astep = a
    m6 = m6 + 1
    Wend
    End If

    gamma = -(TA - firstTA) / (firstTC - DIC3) / 2
    'MsgBox (gamma)
    H3 = lastH
    'MsgBox ("J = " & J & ", m6 = " & m6 & ", lastH = " & lastH)
    If lastH > 0 Then
        pH = -log(lastH) / log(10#)
        Else
        pH = -9999
    End If
    RGasConstant = 83.1451: 'bar-cm3/(mol-K)

    RT = RGasConstant * TempK(1)
    Pbar = P(1) / 100

    'deltaVs are in cm3/mole
    'Kappas are in cm3/mole/bar
    'PROGRAMMER'S NOTE: all logs are log base e
    '
    '

'*********************************************************************

CalculateCa:

    'Riley, J. P. and Tongudai, M., Chemical Geology 2:263-269, 1967:
    calc = 0.02128 / 40.087 * (Sal(1) / 1.80655): ' in mol/kg-SW
    'this is .010285 * Sal / 35
'
'
```

```
CalciteSolubility:
    'Mucci, Alphonso, Amer. J. of Science 283:781-799, 1983.
    logkCa = -171.9065 - 0.077993 * TempK(1) + 2839.319 / TempK(1)
    logkCa = logkCa + 71.595 * lgTempK(1) / log(10!)
    logkCa = logkCa + (-0.77712 + 0.0028426 * TempK(1) + 178.34 /
    TempK(1)) * SqrSal(1)
    logkCa = logkCa - 0.07711 * Sal(1) + 0.0041249 * SqrSal(1) * Sal(1)
    'sd fit = .01 (for Sal part, not part independent of Sal)
    kCa = 10# ^ (logkCa): ' this is in (mol/kg-SW)^2
'
'
AragoniteSolubility:
    'Mucci, Alphonso, Amer. J. of Science 283:781-799, 1983.
    logkAr = -171.945 - 0.077993 * TempK(1) + 2903.293 / TempK(1)
    logkAr = logkAr + 71.595 * lgTempK(1) / log(10)
    logkAr = logkAr + (-0.068393 + 0.0017276 * TempK(1) + 88.135 /
    TempK(1)) * SqrSal(1)
    logkAr = logkAr - 0.10018 * Sal(1) + 0.0059415 * SqrSal(1) * Sal(1)
    'sd fit = .009 (for Sal part, not part independent of Sal)
    kAr = 10# ^ (logkAr): ' this is in (mol/kg-SW)^2
'
'
PressureCorrectionForCalcite:
    'Ingle, Marine Chemistry 3:301-319, 1975
    'same as in Millero, GCA 43:1651-1661, 1979, but Millero, GCA 1995
    'has typos (-.5304, -.3692, and 10^3 for Kappa factor)
    deltaVKCa = -48.76 + 0.5304 * TempC
    KappaKCa = (-11.76 + 0.3692 * TempC) / 1000!
    lnkCafac = (-deltaVKCa + 0.5 * KappaKCa * Pbar) * Pbar / RT
    kCa = kCa * Exp(lnkCafac)
    logkCa = log(kCa) / log(10#)
'
'
PressureCorrectionForAragonite:
    'Millero, Geochemica et Cosmochemica Acta 43:1651-1661, 1979,
    'same as Millero, GCA 1995 except for typos (-.5304, -.3692,
```

```
    'and 10^3 for Kappa factor)
    deltaVKAr = deltaVKCa + 2.8
    KappaKAr = KappaKCa
    lnKArfac = (-deltaVKAr + 0.5 * KappaKAr * Pbar) * Pbar / RT
    kAr = kAr * Exp(lnKArfac)
    logkAr = log(kAr) / log(10#)
'
'
'
'*****************************************************************

CalculateOmegasHere:
    'MsgBox ("H3 = " & H3)
    'MsgBox ("HCO3 = " & HCO3)
    'MsgBox ("CO3 = " & CO3)
    HCO3 = ainit * k1a / H3
    CO3 = k1a * k2a * astep / (H3 * H3)
    DIC3 = ainit + HCO3 + CO3
    CO32 = k2(1) * HCO3 / H3 + ainit * k1(1) * k2(1) / H3 ^ 2 + CO3

    OmegaCa = CO32 * calc / ksp1(1)
    lastOmegaCa = OmegaCa

    OmegaAr = CO32 * calc / ksp2(1)
    lastOmegaAr = OmegaAr
    Range("B150:B14994").Cells(J) = CO32 * calc

'RPOC = (ca1 + ca2) / g3
'DIC3 = DIC3 + lastca1 + lastca2 - ca1 - ca2

'MsgBox ("2 : " & A & ", DIC3 = " & DIC3)
'C = "gamma = " & g3 & ", CO2 = " & A + af3 & "af3 = " & af3
'MsgBox (C)
'Range("AQ150:AQ14994").Cells(J) = g3
'MsgBox ("3: " & A + af3)

'Range("BI150:BI14994").Cells(J) = POC / g3
```

```
    d = BT(1) * H3 / (kB(1) + H3)
    e = BT(1) * kB(1) / (kB(1) + H3)

'MsgBox ("pH = " & pH)

Range("AU150:AU37300").Cells(J) = ainit
Range("AS150:AS37300").Cells(J) = HCO3
Range("AT150:AT37300").Cells(J) = CO3
Range("AN150:AN37300").Cells(J) = H3
Range("AR150:AR37300").Cells(J) = d
Range("AS150:AS37300").Cells(J) = e
Range("BJ150:BJ37300").Cells(J) = OmegaCa
Range("BK150:BK37300").Cells(J) = 0 '= ca2
Range("AQ150:AQ37300").Cells(J) = TA
Range("AO150:AO37300").Cells(J) = pH
Range("BO150:BO37300").Cells(J) = fCo2(1)
Range("BK150:BK37300").Cells(J) = OmegaAr
Range("BQ150:BQ37300").Cells(J) = k0(1) * fCo2(1) * Av2
Range("AT150:AT37300").Cells(J) = gamma
'Range("AP150:AP14994").Cells(J) = a + HCO3 + CO3
H3 = lastH
'MsgBox (lastH)
DIC2 = DIC3
'MsgBox ("Revelle = " & RFgamma)

    If logflag = 1 Or bflag = 1 Or specflag = 1 Then
        Close #FileNum
    End If
        J = J + 1
    Wend

ActiveWorkbook.ActiveSheet.Calculate
For z1 = 1 To 10000000
Next

Range("A32:A150").Cells(count) = Range("Q26:Q26").Cells(1)
Range("I32:I150").Cells(count) = Range("Q24:Q24").Cells(1)
```

```
Range("G32:G150").Cells(count) = Range("R24:R24").Cells(1)
Range("P32:P150").Cells(count) = Range("P19:P19").Cells(1)
Range("Q32:Q150").Cells(count) = Range("P21:P21").Cells(1)
Range("H32:H150").Cells(count) = Range("R27:R27").Cells(1)
Range("R32:R150").Cells(count) = Range("BJ146:BJ146").Cells(1)
Range("S32:S150").Cells(count) = Range("BK146:BK146").Cells(1)
Range("Q32:Q150").Cells(count) = Revelle

Range("P32:P150").Cells(count) = Range("AT145:AT145").Cells(1)
    If firstpass2 = 1 Then
        sumval1 = Range("AF141:AF141").Cells(1)
        ffac = 1
        'MsgBox ("ffac = " & ffac)
    ElseIf firstpass2 = 2 Then
        sumval2 = Range("AF141:AF141").Cells(1)
    End If
    Range("U32:U150").Cells(count) = Av2
    N1 = 10150 - 150 + 1

    If firstpass2 = 1 Then
        count = 2006 - 1990 + 1
        firstpass2 = 2
    ElseIf firstpass2 = 2 Then
        count = 1
        firstpass2 = 0
    Else
        count = count + 1
    End If

Wend

'ActiveWorkbook.Save

End

    Error = 0
    On Error Resume Next
```

```
    If Error Then
        Close #FileNum
        MsgBox (Error)
    Return
    End If

    End Sub

Public Function Sgn1(ByRef x As Double)
    If x < 0 Then
        Sgn1 = -1
    Else
        Sgn1 = 1
    End If

End Function

Public Function Sqr(ByRef x As Double) As Double
Sqr = x * x
End Function

Public Function Sqrt(ByRef x As Double)
Dim Temp As Double, Temp2 As Double

    If x = 0# Then
        Sqrt = 0#
        Exit Function
    End If

Temp = 0.5 * log(x)
Temp2 = Exp(Temp)
Sqrt = Temp2
End Function

Public Function Atan2(ByRef x As Double, ByRef y As Double)
Dim Temp1 As Double, Temp2 As Double, Hyp As Double
Dim Sgn1 As Double, Sgn2 As Double
```

```
Dim Angle As Double, PI As Double

PI = 3.14159265358979

Temp1 = Abs(x)
Temp2 = Abs(y)
    If x < 0# Then
        Sgn1 = -1#
    Else
        Sgn1 = 1#
    End If
If y < 0# Then
        Sgn2 = -1#
    Else
        Sgn2 = 1#
    End If

    Hyp = Sqrt(Temp1 * Temp1 + Temp2 * Temp2)
    If Hyp = 0# Then
        Angle = 0#
    ElseIf Temp1 = 0# And Temp2 > 0# Then
        Angle = PI / 2#
    ElseIf Temp2 = 0# And Temp1 > 0# Then
        Angle = 0#
    Else
        Angle = Acos(Temp1 / Hyp)
    End If
    If Sgn1 < 0 And Sgn2 < 0 Then
        Angle = Angle + PI
    ElseIf Sgn1 < 0 And Sgn2 > 0 Then
        Angle = PI - Angle
    ElseIf Sgn1 > 0 And Sgn2 < 0 Then
        Angle = 2# * PI - Angle
    End If

    Atan2 = Angle
```

```
End Function
Public Function Acos(ByRef Cos1 As Double)
Dim Theta1 As Double, Theta2 As Double, Theta3 As Double, f1 As
Double, f2 As Double, f3 As Double
Dim limit As Double
Dim Temp As Double
Dim flag1 As Boolean
Dim flag2 As Boolean
Dim flag3 As Boolean
Dim SMALL As Double
Dim PI As Double
Dim flag4 As Boolean

SMALL = 1e-15
PI = 3.14159265358979

    If Abs(Cos1 - 1#) <= 0.0001 Then
        Acos = 0#
    Exit Function
    ElseIf Abs(Cos1) < 0.0001 Then
        Acos = PI / 2#
    Exit Function
    End If

    If Cos1 > 0.9 Then
        Theta1 = 0.3176
    ElseIf Cos1 > 0.8 Then
        Theta1 = 0.5548
    ElseIf Cos1 > 0.7 Then
        Theta1 = 0.7227
    ElseIf Cos1 > 0.6 Then
        Theta1 = 0.8632
    ElseIf Cos1 > 0.5 Then
        Theta1 = 0.9884
    ElseIf Cos1 > 0.4 Then
        Theta1 = 1.104
```

```
ElseIf Cos1 > 0.3 Then
    Theta1 = 1.213
ElseIf Cos1 > 0.2 Then
    Theta1 = 1.318
ElseIf Cos1 > 0.1 Then
    Theta1 = 1.42
Else
    Theta1 = 1.521
End If

f1 = Calcf(Theta1, Cos1)
Theta2 = Theta1 * 1.005
f2 = Calcf(Theta2, Cos1)
limit = 10000#
Theta3 = Theta2 - f2 * (Theta2 - Theta1) / (f2 - f1)
If Theta3 < 0# Then
    MsgBox ("Theta3 < 0")
    MsgBox (f1)
    MsgBox (f2)
    Theta3 = Theta2 - 0.3 * f2 * (Theta2 - Theta1) / (f2 - f1)
End If

f3 = Calcf(Theta3, Cos1)
Theta1 = Theta2
Theta2 = Theta3
f1 = f2
f2 = f3
Temp = 1# - Theta2 / Theta1
Temp = Abs(Temp)
flag1 = Temp > 1e-05
flag2 = limit > 0#
flag3 = Abs(f1) > SMALL And Abs(f2) > SMALL And Abs(f2 - f1) >
SMALL * SMALL
flag4 = Abs(Theta2) > 0.0001
While (flag1 And flag2 And flag3 And flag4)
    Theta3 = Theta2 - f2 * (Theta2 - Theta1) / (f2 - f1)
```

```
If Theta3 < 0# Then
    MsgBox ("Theta3 < 0")
    MsgBox (Theta1)
    MsgBox (Theta2)
    MsgBox (f1)
    MsgBox (f2)
    Theta3 = Theta2 - 0.3 * f2 * (Theta2 - Theta1) / (f2 - f1)
End If

f3 = Calcf(Theta3, Cos1)
Theta1 = Theta2
Theta2 = Theta3
f1 = f2
f2 = f3
limit = limit - 1#
Temp = Abs(1# - Theta2 / Theta1)
flag1 = Temp > 1e-05
flag2 = limit > 0#
flag3 = Abs(f1) > SMALL And Abs(f2) > SMALL And Abs(f2 - f1) >
SMALL * SMALL
flag4 = Abs(Theta2) > 0.0001
Wend

Acos = Theta2

End Function

Public Function Calcf(ByRef Theta As Double, ByRef Cos1 As Double)
Dim Z As Double

If Theta < 0# Then
    MsgBox ("theta < 0")
ElseIf Theta = 0# Then
    Z = -10000000#
Else
    Z = log(Theta) / log(Exp(1#))
End If
```

```
        Calcf = 1 - Exp(2# * Z) / 2# + Exp(4# * Z) / 24# - Exp(6# * Z)
        / 720# + Exp(8# * Z) / 40320# - Exp(10# * Z) / 3628800# +
        Exp(12# * Z) / 479001600# - Exp(14# * Z) / 87178291200# - Cos1
    End Function

Public Function sort(ByRef data() As Double, ByRef count As Long)
Dim I As Long, J As Long
Dim Temp As Double
Dim intpart As Long

intpart = floor(0.225 * count + 1#)

    For I = 1 To intpart
    Temp = data(I)
    For J = I + 1 To count
    If data(J) > Temp Then
        data(I) = data(J)
        data(J) = Temp
        Temp = data(I)
    End If
    Next
    Next

    sort = intpart

    End Function

Public Function Round(x As Double) As Long
Dim y As Double, L As Long, K As Long

    K = floor(x)
    If x < 0 Then
        L = floor(x - 0.5)
    ElseIf x > 0 Then
        L = floor(x + 0.5)
    Else
        L = 0
```

```
    End If

    If Abs(L) > Abs(K) Then
        Round = L
    Else
        Round = K
    End If

    End Function

Public Function floor(x As Double) As Long

Dim J As Long

    J = x
    If x > 0 Then
    If J > x Then
        J = J - 1
    End If
    ElseIf x < 0 Then
    If J < x Then
        J = J + 1
    End If
    End If

    floor = J

    End Function

Sub LogInformation(LogMessage As String)
Const LogFileName As String = "excellogfile1.txt"
Dim FileNum As Integer
FileNum = FreeFile ' next file number
Open LogFileName For Append As #FileNum ' creates the file if it doesn't exist
Print #FileNum, LogMessage ' write information at the end of the text file
Close #FileNum ' close the file
End Sub
```

```
Sub DeleteLogFile()

On Error Resume Next ' ignore possible errors
Kill "excellogfile1.txt" ' delete the file if it exists and it is possible
On Error Resume Next
If Error Then MsgBox (Error)
End Sub
```

Appendix C - Program to calculate Earthquake probabilities into Future

Earthquake probability Distribution code

(* COPYRIGHT 2012,2013,2014,2015,2016,2017
CHONDRALLY/PSYREIGHE, ASA, AGU, IEEE, AAAS, ACS, ACM
ALL RIGHTS RESERVED)
(* Forecasting probability, magnitude and time of occurrence \
software using time series and Density Functional Bayesian Martingale \
Monte Carlo Path Integration and Fourer Extrapolation.
This software can Forecast the likely times and magnitudes of \
earthquakes or any events with a magnitude and a time stamp. It can \
be used to forecast terrorist strikes if a past time series of \
magnitudes and timestamps is known for certain regions or cells. It \
can be used to forecast losses or wins in military campaigns. it can \
be used to forecast emergency calls if they are ranked in magnitude \
with a time stamp. it can be used to forecast suicide calls or \
distress calls. it can be used to forecast crimes and murders and \
robberies if a magnitude (category) is assigned with a time stamp. \
It can help solve serial killings and serial crimes and drug deals. \
it can be used in another form to forecast the stock market or stock \
market crashes or surges. it can be used to forecast extreme events \
of all kinds like hurricanes and tsunamis and volcanic eruptions. It \
can be used to forecast traffic jams and traffic accidents or any \
rare events like hijackings or airplane accidents or events like 911 \

or the occurrence of declarations of war between nations or more \
importantly the occurrence of trade deals or peace treaties. it can \
be used to forecast the outbreak of riots or rebellions or the \
occurrence of music concerts and the generation of new pop - stars. \
It can be used in epidemiology to forecast the outbreak of diseases \
in many regions of the world and outbreaks of manias, panics and \
crashes.
References:
Density Functional Stock Forecasting and American Option pricing \
compared to Black-Scholes(BS) using Bayesian Markov Monte Carlo \
Simulation and Wavelet or Fourier or Neural Network Extrapolation \
with Indicators.
http://library.wolfram.com/infocenter/MathSource/9086/

Density Functional American Option pricing with Bayesian Monte Carlo \
Path Int& MUSIC w/Kelly Crit
https://www.mathworks.com/matlabcentral/fileexchange/56352-density-\
functional-american-option-pricing-with-bayesian-monte-carlo-path-int—\
music-w—kelly-crit

Would the magnesium carbonate buffer in the ocean break as CO2 \
increases, When?
https://www.thenakedscientists.com/forum/index.php?topic=53181.\
msg452992#msg452992

Can we build an efficient hybrid solar-natural gas engine that emits \
no CO2?
https://www.thenakedscientists.com/forum/index.php?topic=53180.\
msg447667#msg447667
*)

```
Needs["Calendar`"]

ClearAll[matlabFind1]
matlabFind1[lst_List,
Op_: {Unequal, Greater, Equal, GreaterEqual, LessEqual},
```

```
elt_: 0] := Module[{x}, Flatten[Position[lst, x_ /; Op[x, elt]]]];

ClearAll[MusicFinal];
MusicFinal[x_List, futuresteps_Integer, s0_Real, N1_Integer,
flatflag_Integer, debugflag_Integer] :=
Module[{N1a, coeffs, deTrendedData, volatility, x2, p, lags1,
sum1, lags2, R2, R, R1, J, eigvecs, eigs, eqn1, roots2, theroots2,
w, w2, Amps, theta, s, w5, i, j, count, N2, w3, w4, magphase, y,
plot1, plot2, delk, m, k, magphase2, aminus, v1, v2, z1, coeffs2,
temp1, aplus, G, S1, S2, indx, indx2, roots1, z, pi, y1, y3, t1,
t2, t3, t4, temp, x1, X, temp2, temp0, astar, ntemp, meanval,
meanx},

If[debugflag == 1, Print["**1"]];
N2 = Length[x];
N1a = N1;

If [N1a > N2 - 1, N1a = N2 - 1];
If[N1a > 100, N1a = 100];
If[debugflag == 1, Print["N2=", N2]];
If[debugflag == 1, Print["x=", x];];
Clear[coeffs, deTrendedData, volatility];
If [flatflag == 0,
{coeffs, deTrendedData, volatility} =
getPercentErrorLinear[x, debugflag];
,
coeffs = ConstantArray[0.0, 2];
coeffs = 0;
coeffs = Mean[x];
deTrendedData = x - coeffs;
];

x2 = deTrendedData;

N2 = Length[deTrendedData];

p = 3;
```

```
If[debugflag == 1, Print["**1a"]];
lags1 = lags2 = ConstantArray[0.0, N1a];
Do[If[i == 0, sum1 = 0;
Do[sum1 = sum1 + x2*x2;, {j, 1, N2}];
lags1 = N[sum1/(N2 - 1)]; lags2 = N[sum1/(N2 - 1)];
, sum1 = 0;
Do[sum1 = sum1 + x2*x2, {j, 1, N2 - i}];
lags1 = N[sum1/(N2 - i - 1)]; sum1 = 0;
Do[sum1 = sum1 + x2*x2;, {j, N2, i + 1, -1}];
lags2 = N[sum1/(N2 - i - 1)];],
{i, 0, N1a - 1}];
If[debugflag == 1, Print["**1a1"]];
R1 = R2 = R = ConstantArray[0, {N1a, N1a}];
Do[R1 = lags1;
R2 = lags2;, {i, 1, N1a}, {j, 1, N1a}];
J = ConstantArray[0, {N1a, N1a}];
Do[J = 1;, {i, 1, N1a}];
If[debugflag == 1, Print["**1a2"]];
R1 = (R1 + J.Transpose[R1].J)/2;
R2 = (R2 + J.Transpose[R2].J)/2;
R = (R1 + R2)/2;
R = (R + Transpose[R])/2;
eigvecs = Eigenvectors[R];
If[debugflag == 1, Print["**1b"]];
eigs = Eigenvalues[R];
m = Ordering[eigs];
eqn1 = 0;
Clear[s]
Clear[aminus, aplus]; aminus = aplus = ConstantArray[0, {1, N1a}];
Do[aminus = z^(-(j - 1));
aplus = z^(j - 1);, {j, 1, N1a}];
Clear[G];
G = ConstantArray[0, {N1a, N1a}];
S1 = Conjugate[Transpose[eigvecs]];
S2 = eigvecs;
G = S1.S2;
```

```
eqn1 = aminus.G.Transpose[aplus];

If[debugflag == 1, Print["**1c"]];

Clear[roots2, roots1, theroots2]; theroots2 = {};
Off[Solve::ratnz];
roots2 = z /. Solve[eqn1 == 0, z];
theroots2 = Append[theroots2, roots2];
If[debugflag == 1, Print["**1d"]];
theroots2 = Flatten[Flatten[theroots2]];
indx2 = matlabFind1[Abs[theroots2], Less, 1];

roots1 = theroots2]];
If[debugflag == 1, Print["roots=", roots1]];
w = ArcTan[Re[roots1], Im[roots1]];
If [debugflag == 1, Print["step2"]];
Clear[w2]; w2 = {}; pi = 4*ArcTan[1];
Do[If[w >= 0 && w < pi, w2 = Append[w2, w];], {i, 1,
Length[w]}];(* select low frequency spectrum *)
astar = Floor[2^(Floor[Log[N2]/Log[2]] + 2)/8192 + .5];
If[astar < 1, astar = 1]; ntemp = 8192*astar;
Print["astar = ", astar, ", ntemp = ", ntemp];
temp = ConstantArray[0.0, ntemp];
temp = x;
X = ForwardFFT[temp];
If [debugflag == 1, Print["step3"]];
(*Clear[temp1];
temp1 = ConstantArray[{},512-10+1];
Do[temp1={i+9,Abs[X]},{i,1,512-10+1}];
plot1 = ListLinePlot[temp1,PlotRange\[Rule] Automatic,
PlotStyle\[Rule] Red,PlotLabel\[Rule] "Fourier Transform ABS",
Joined\[Rule] True];*)
temp2 = Max[Abs[X]];
temp0 = matlabFind1[Abs[X], Equal, temp2];
If[Length[temp0] > 0,
If[debugflag == 1, Print["temp0=", temp0 + astar*53]];
```

```
temp = N[(temp0 + astar*53)*2/ntemp*pi];
If[debugflag == 1,
Print["rad ratio = ", N[astar*4096/(temp0 + astar*53)]]];
w2 = Append[w2, temp];
];
temp2 =
Max[Abs[X + astar*53 + astar*16 ;; astar*300]]]];
temp0 = matlabFind1[Abs[X], Equal, temp2];
If[Length[temp0] > 0,
temp = N[(temp0 + astar*53)*2/ntemp*pi];
w2 = Append[w2, temp];
];
temp2 =
Max[Abs[X + astar*53 + astar*16 ;; astar*350]]]];
temp0 = matlabFind1[Abs[X], Equal, temp2];
If[Length[temp0] > 0,
temp = N[(temp0 + astar*53)*2/ntemp*pi];
w2 = Append[w2, temp];
];
temp2 =
Max[Abs[X + astar*53 + astar*16 ;; astar*400]]]];
temp0 = matlabFind1[Abs[X], Equal, temp2];
If[Length[temp0] > 0,
temp = N[(temp0 + astar*53)*2/ntemp*pi];
w2 = Append[w2, temp];
];
temp2 =
Max[Abs[X + astar*53 + astar*16 ;; astar*450]]]];
temp0 = matlabFind1[Abs[X], Equal, temp2];
If[Length[temp0] > 0,
temp = N[(temp0 + astar*53)*2/ntemp*pi];
w2 = Append[w2, temp];
];
w3 = Sort[w2];
w4 = {};
i = 1;
```

```
While[i <= Length[w3],
k = w3;
i = i + 1;
While[i <= Length[w3] && w3 == k,
i = i + 1;
];
w4 = Append[w4, k];
];

Clear[Amps, theta, y3, delk];
If[debugflag == 1, Print["w4=", w4]];

If[debugflag == 1, Print["**1e"]];
delk = N[1/2/pi];
y3 = ConstantArray[0, {2, N2 + futuresteps}];
Amps = magphase = theta = ConstantArray[0, Length[w4]];
Do[magphase =
magphase + x2*Exp[-Sqrt[-1]*w4*(j - 1)];, {j, 1,
N2}, {i, 1, Length[w4]}];

magphase = magphase/N2;

Do[Amps = Abs[magphase];
theta = ArcTan[Re[magphase], Im[magphase]], {i, 1,
Length[w4]}];
If[debugflag == 1, Print["**1f"]];
y = ConstantArray[0, N2 + futuresteps];
Do[y =
y + 2*Amps*Cos[w4*(j - 1) + theta];, {j, 1,
N2 + futuresteps}, {i, 1, Length[w4]}];

Do[y3 = y;, {i, 1, N2 + futuresteps}, {m, 1, 2}];
If [debugflag == 1, Print["*3"];];
Do[magphase = 0;
Do[magphase =
magphase +
x2*Exp[Sqrt[-1]*w4*(1 - delk*k)*(j - 1)];, {j, 1,
```

```
N2}];
Amps = Abs[magphase];
theta = ArcTan[Re[magphase], Im[magphase]];
Do[y3, j]] =
y3, j]] +
2*Amps*
Cos[w4*(1 - delk*k)*(j - 1) + theta];, {j, N2 + 1,
N2 + futuresteps}], {i, 1, Length[w4]}, {k, -1, 1, 2}];
If[debugflag == 1, Print["**1f1", y3]];

If[debugflag == 1, Print["**1g"]];
Clear[t1, t2, t3, t4, temp];
y1 = ConstantArray[0.0, N2 + futuresteps];
x1 = ConstantArray[0.0, N2];
temp = N[Mean[y]];
y1 = y - temp;
x1 = x2 - N[Mean[x2]];
v1 = ConstantArray[0.0, N2 + futuresteps];
Clear[coeffs2];
{plot1, coeffs2, v1} = getPercentErrorLinear3[x1, y1, debugflag];
Clear[y];
y = ConstantArray[0, {3, N2 + futuresteps}];
y = v1;
If[debugflag == 1, Print["*1"];];
y = y;
y = y;
If[debugflag == 1, Print["*2"];];
y =
N[coeffs2*(y3 - temp) + temp];
y =
N[coeffs2*(y3 - temp) + temp];
If[debugflag == 1, Print["v1=", v1];];

meanval = ConstantArray[0, 3];
meanval = Mean[y];
meanval = Mean[y];
```

```
meanval = Mean[y];

meanx = Mean[x];

Do[y =
y + coeffs*i - meanval + meanx;, {i, N2 + 1,
N2 + futuresteps}, {k, 1, 3}];

{plot1, y}];

ClearAll[getPercentErrorLinear] (* definition from Feb 7, 2013
where actdata is original stock prices, dataarray is either Null \
matrix {{}} or temperatures SeasData *)

getPercentErrorLinear[actdata_List, debugflag_Integer] :=
Module[{delta, i, j, k, a = 1, b, N1, N2, R, theta, coeffs, vol6,
vol7, x5, y5, L1, L, Temp, temp0, temp1, temp2, Soy, Soy2, Sxoy,
Sxoy2, Sx2oy2, vol, volatility, vol2, deTrendedData}, N1 = 0;
N2 = Length[actdata]; L = N2 - N1; L1 = N1 + 1;
coeffs = ConstantArray[0, 2]; vol6 = ConstantArray[0, L]; vol7 = 0;

Temp = N[SetPrecision[actdata, 60] + 1000.0];
temp0 = matlabFind1[Abs[Temp] LessEqual, 10^(-60)];
If[Length[temp0] > 0, Temp = 0.0];
temp1 = matlabFind1[Temp, Unequal, 0.0];
If[debugflag == 1, Print["***1a"]];
N2 = L = Length[Temp];
L1 = 1;
temp2 = Temp;
L = Length[temp2];
Soy = Total[1/(10^(-60) + temp2)];
Soy2 = Total[1/(10^(-60) + temp2^2)];
Sxoy = Total[temp1/(10^(-60) + temp2)];
Sxoy2 = Total[temp1/(10^(-60) + temp2^2)];
Sx2oy2 = Total[temp1^2/(10^(-60) + temp2^2)];
If[debugflag == 1, Print["***1b"]];
N2 = Length[temp1];
```

```
If[debugflag == 1, Print["Sxoy2=", Sxoy2];
Print["Soy=", Soy, ", Sxoy=", Sxoy, ", Soy2=", Soy2];
Print["Sxoy2 =", Sxoy2, ", Sx2oy2 =", Sx2oy2, ", Soy2 =", Soy2];];
coeffs =
N[(Sxoy2*Soy - Sxoy*Soy2)/(10^(-60) + Sxoy2^2 - Sx2oy2*Soy2)];
If[debugflag == 1, Print["***1ba"]];
coeffs = N[(Soy - coeffs*Sxoy2)/(10^(-60) + Soy2)];
If[debugflag == 1, Print["***1c"]];
L = Length[Temp];
vol = ConstantArray[0, L];
deTrendedData = ConstantArray[0, L];
vol7 = 0;
Do[

deTrendedData =
temp2 - coeffs*temp1 - coeffs;
vol = (deTrendedData)^2;
vol7 = vol7 + vol, {j, 1, L}];
If[debugflag == 1, Print["***1d"]];

coeffs = coeffs - 1000.0;
{coeffs, deTrendedData, vol7}];

ClearAll[getPercentErrorExp] (* definition from Feb 7, 2013
where actdata is original stock prices, dataarray is either Null \
matrix {{}} or temperatures SeasData *)

getPercentErrorExp[actdata_List, debugflag_Integer] :=
Module[{delta, i, j, k, N1, N2, L1, L, a1, b1, c1, a2, b2, c2, a3,
b3, c3, a, b, c, fc1, fc2, fc3, x, temp, temp0, temp1, temp2,
temp3, temp4, temp5, Soy2, Soy3, Sexpoy2, Sexpoy3, Sxexpoy2,
Sexp2oy3, Sxexpoy3, Sxexp2oy3, flag},

temp2 = N[SetPrecision[Transpose[actdata], 60]];
temp1 = Transpose[actdata];
temp3 = matlabFind1[Abs[temp2], GreaterEqual, 10^(-200)];
temp4 = temp2;
```

```
temp5 = temp1;
If[debugflag == 1, Print["***1a"]];
c = .0001;
Soy2 = Total[N[1/(temp4^2)]];
Soy3 = Total[N[1/(temp4^3)]];
Sexpoy2 = Total[N[Exp[-c*temp5]/(temp4^2)]];
Sexpoy3 = Total[N[Exp[-c*temp5]/(temp4^3)]];
Sexp2oy3 = Total[N[Exp[-2*c*temp5]/(temp4^3)]];
Sxexpoy2 =
Total[N[temp5*Exp[-c*temp5]/(temp4^2)]];
Sxexpoy3 =
Total[N[temp5*Exp[-c*temp5]/(temp4^3)]];
Sxexp2oy3 =
Total[N[temp5*Exp[-2*c*temp5]/(temp4^3)]];
Clear[temp, temp1];
temp =
Solve[-Sexpoy2 + a*Sexpoy3 + b*Sexp2oy3 ==
0 && -Soy2 + a*Soy3 + b*Sexpoy3 == 0, {a, b}];
{a1, b1} = Flatten[{a, b} /. temp];
c1 = c;

If[debugflag == 1, Print["***1b", a1, b1, c1]];

fc1 = Sxexpoy2 - a1*Sxexpoy3 - b1*Sxexp2oy3;
c = .0002;
Soy2 = Total[N[1/(temp4^2)]];
Soy3 = Total[N[1/(temp4^3)]];
Sexpoy2 = Total[N[Exp[-c*temp5]/(temp4^2)]];
Sexpoy3 = Total[N[Exp[-c*temp5]/(temp4^3)]];
Sexp2oy3 = Total[N[Exp[-2*c*temp5]/(temp4^3)]];
Sxexpoy2 =
Total[N[temp5*Exp[-c*temp5]/(temp4^2)]];
Sxexpoy3 =
Total[N[temp5*Exp[-c*temp5]/(temp4^3)]];
Sxexp2oy3 =
Total[N[temp5*Exp[-2*c*temp5]/(temp4^3)]];
```

```
Clear[temp, temp1];
temp =
Solve[-Sexpoy2 + a*Sexpoy3 + b*Sexp2oy3 ==
0 && -Soy2 + a*Soy3 + b*Sexpoy3 == 0, {a, b}];
{a2, b2} = Flatten[{a, b} /. temp];
Clear[a, b, temp2, temp3];
(*a=Range[500,1500,1];
temp1 = -Soy2/Sexpoy3+Soy3/Sexpoy3*a;
temp2 = Sexpoy2/Sexp2oy3+
plot1 = ListLinePlot[*)
c2 = c;
If[debugflag == 1, Print["***1c", a2, b2, c2]];

fc2 = Sxexpoy2 - a2*Sxexpoy3 - b2*Sxexp2oy3;
flag = 0;
While [
flag == 0 && (Abs[1 - c2/c1] > 10^(-6) ||
Abs[1 - b2/b1] > 10^(-6) || Abs[1 - a2/a1] > 10^(-6)),
If [fc2 != 0 && (fc2 - fc1) != 0,
c3 = N[c2 - (c2 - c1)/(fc2 - fc1)];
c = c3;

Soy2 = Total[N[1/(temp4^2)]];
Soy3 = Total[N[1/(temp4^3)]];
Sexpoy2 = Total[N[Exp[-c*temp5]/(temp4^2)]];
Sexpoy3 = Total[N[Exp[-c*temp5]/(temp4^3)]];
Sexp2oy3 = Total[N[Exp[-2*c*temp5]/(temp4^3)]];
Sxexpoy2 =
Total[N[temp5*Exp[-c*temp5]/(temp4^2)]];
Sxexpoy3 =
Total[N[temp5*Exp[-c*temp5]/(temp4^3)]];
Sxexp2oy3 =
Total[N[temp5*Exp[-2*c*temp5]/(temp4^3)]];
Clear[temp, temp1];
temp =
Solve[-Sexpoy2 + a*Sexpoy3 + b*Sexp2oy3 ==
```

```
0 && -Soy2 + a*Soy3 + b*Sexpoy3 == 0, {a, b}];
{a3, b3} = Flatten[{a, b} /. temp];
If[debugflag == 1, Print["a= ", a3, ", b= ", b3, ", c= ", c]];
fc3 = Sxexpoy2 - a3*Sxexpoy3 - b3*Sxexp2oy3;

fc1 = fc2;
fc2 = fc3;
a1 = a2;
a2 = a3;
b1 = b2;
b2 = b3;
c2 = c3;
,
flag = 1;
];
];
a = a2;
b = b2;
c = c2;
{a, b, c}];

ClearAll[getPercentErrorLinear3] (* definition from Feb 7, 2013
where actdata is original stock prices, dataarray is either Null \
matrix {{}} or temperatures SeasData *)

getPercentErrorLinear3[x_List, y_List, debugflag_Integer] :=
Module[{z, vol1, delta, futuresteps, f1, f2, f3, i, j, k, a, b, N1,
N2, R, theta, coeffs, vol6, vol7, x5, y5, L1, L, a2, a1, Temp,
temp0, Soy, Soy2, Sxoy, Sxoy2, Sx2oy2, vol, volatility, vol2,
deTrendedData, x1, y1, MostPurseAir, coeffs3, xvals, vol9, m,
plot1, plot2, t1, t2, t3, t4, t5, t1a, t2a, t3a, t4a, temp1,
temp2, temp3, temp4, temp5, temp1a, temp2a, temp3a, temp4a, t0,
t0a, temp0a}, N1 = 0; N2 = Length[x]; L = N2;
futuresteps = Length[y] - N2;
coeffs = ConstantArray[0, 3];
vol1 = ConstantArray[0, 3];
```

```
a = ConstantArray[0, 3];
a = 1;
a = .95;

k = 3;
x1 = x - Mean[x];
y1 = y - Mean[y];
t0 = Max[x1]];
t0a = Min[x1]];
t1 = Max[x1]];
t1a = Min[x1]];
t2 = Max[x1]];
t2a = Min[x1]];
t3 = Max[x1]];
t3a = Min[x1]];
t4 = Max[x1];
t4a = Min[x1];
t5 = Abs[x1];
(*temp1=matlabFind1[Abs[x]],Equal,t1];
temp2 = matlabFind1[Abs[x]],Equal,
t2];
temp3 =matlabFind1[Abs[x]],Equal,
t3];
temp4=matlabFind1[Abs[x],Equal,t4];
temp5=N2;*)
temp0 = Max[y1]];
temp0a = Min[y1]];
temp1 = Max[y1]];
temp1a = Min[y1]];
temp2 = Max[y1]];
temp2a = Min[y1]];
temp3 = Max[y1]];
temp3a = Min[y1]];
temp4 = Max[y1];
temp4a = Min[y1];
```

```
temp5 = Abs[y1];
a =
N[((t0 - t0a)/(temp0 - temp0a) + (t1 - t1a)/(temp1 -
temp1a) + (t2 - t2a)/(temp2 - temp2a) +
2*(t3 - t3a)/(temp3 - temp3a) +
2*(t4 - t4a)/(temp4 - temp4a))/7];

Clear[MostPurseAir];
MostPurseAir = ConstantArray[0, N2 + futuresteps];
Do[MostPurseAir = a*y1 + Mean[y];, {i, 1,
N2 + futuresteps}];
If[debugflag == 1, Print["MostPurseAir=", MostPurseAir];
Print["x1=", x - Mean[x]];];
Clear[temp1];
temp1 =
StringJoin["MostPurseAir vs. x, a=", ToString[N[a]]];

plot1 =
ListLinePlot[{MostPurseAir, x1},
PlotRange -> Automatic, PlotLabel -> temp1,
PlotStyle -> {Red, Black}, GridLines -> Automatic,
Joined -> True];

{plot1, a, MostPurseAir}];

ClearAll[ForwardFFT];
ForwardFFT[x_List] := Module[{N1, odd, t1, t2, even, W, X, k, t},
N1 = Length[x];

X = ConstantArray[0.0, N1];
If [N1 > 1,
odd = ConstantArray[0.0, N1/2];
even = ConstantArray[0.0, N1/2];
odd = x;
even = x;
t1 = ForwardFFT[odd];
```

```
t2 = ForwardFFT[even];
X = Join[t2, t1];

Do[W = Exp[-Sqrt[-1]*N[2*Pi*(k - 1)/N1]]; t = X;
X = t + W*X;
X = t - W*X, {k, 1, N1/2}]
,
X = {x}];
X]

ClearAll[getData];
getData[data_List, numpts_Integer: 8192, extendflag_,
debugflag_Integer] /; (debugflag == 1 || debugflag == 0) :=
Module[{xa, xb, MaxRatio, ratios, difflist, phaselist, extenlst,
temp1, intlist, data1, pdflist, outlst, sdifflist, N2, k1, k, j},
If[debugflag == 1, Print["Start getData:"]; Print[Length[data]]];
temp1 = matlabFind1[data, Greater, 0];
MaxRatio = N[Max[data]/(10^(-200) + Min[data])];

If[debugflag == 1, Print[Sort[data]]];
Clear[ratios];
ratios = SetPrecision[ConstantArray[0, Length[temp1] - 1], 30];
j = 1; k1 = 1;
While[j < Length[temp1],
While[j < Length[temp1] && Abs[data]]] == 0,
j = j + 1]; k = j + 1;
While[k <= Length[temp1] && Abs[data]]] == 0,
k = k + 1;];
ratios = N[data]]/data]]];
k1 = k1 + 1; j = j + 1;];
N2 = 2^(Floor[Log[Length[ratios]/Log[2.0]] + 1]);
If [N2 < numpts, N2 = numpts];
If[debugflag == 1, Print["ratios=", ratios]; Print["3a"]];
If [Max[Abs[ratios]] >
0, {xa, xb, extenlst, intlist, pdflist, outlst} =
getPDF5[data, N2, extendflag, debugflag];, xa = xb = 0;
```

```
extenlst =
intlist = pdflist = outlst = ConstantArray[0, Length[ratios]];];
If[debugflag == 1, Print["3b"]];
{xa, xb, data, extenlst, intlist, pdflist, outlst}]

ClearAll[getPDF5]; (* new definition from 8/10/2017 *)

getPDF5[temp_List, N2_Integer, extendflag_, debugflag_Integer: 0] :=
Module[{sortlst, N1, xa, xb, extenlst, extenlst1, i, h0, intlist,
outlst, h7, coeffs, FlagRatioa2, FlagRatiob1, j, x0, sumval, m,
mat, y, x, z, a, b, x1, x2, c, d, a5, b5, c5, d5, a7, b7, c7, d7,
nlm1, a2, f7, c2, d2, f1, nlm2, b1, f8, c1, d1, z1, f2, count,
temp1, n, delta, f11, f12, k, flag1, down, up, order, n1, n2, N3,
N5, flag, intpdflist, pdflist, densitylist, density, y2, a3, b2,
temp2, sumval1, func1, pdf, int, f, temp3, z2, k1, pdflist2, h01,
nlm1pars, nlm2pars, a4, b4, pdlist, N7},
If [debugflag == 1, Print["Start GetPDF5"]];
sortlst = Sort[temp]; N1 = Length[temp];
xa = sortlst; xb = sortlst; extenlst1 = sortlst;
extenlst = {};

flag = 0;
temp1 = matlabFind1[extenlst1, Greater, 0];
extenlst1 = Sort[extenlst1];
N1 = Length[extenlst1];
Do[If[extenlst1 <= 0 || extenlst1 == Infinity ||
extenlst1 != NumberQ, flag = 1;];, {i, 1, N1}];
If[flag == 1,
Print["GetPDF5:Illegal: Numbers in the list are either not \
positive, zero, infinity or not a number:", Sort[extenlst1]]; Exit[];];
extenlst = {}; pdlist = {};
j = 1; While [j < N1, k = 1;
While[j < N1 && extenlst1 == extenlst1, k = k + 1;
j = j + 1;]; AppendTo[pdlist, k];
AppendTo[extenlst, extenlst1]; j = j + 1];
If[extenlst1 != extenlst1, AppendTo[pdlist, 1];
```

```
AppendTo[extenlst, extenlst1];];
N1 = Length[pdlist];
If[N1 != Length[extenlst],
Print["pdlist is not same length as extenlst: ERROR"]; Exit[]];

N5 = 2^(Floor[Log[N1]/Log[2] + 1]);
If [N5 < N2, N5 = N2];

xa = extenlst; xb = extenlst;

If[extendflag > 0,
a = xa;
b = xb;
If[debugflag == 1, Print["4a"];];
Print[extenlst];
nlm1pars[a2, f7, c2, d2, x] =
Log[Abs[a2]] + c2*Log[x + 2. + f7] + d2*Log[x + 2. + f7]^2;
Off[NonlinearModelFit::nrlnum, NonlinearModelFit::cvmit,
NonlinearModelFit::sszero];
If[debugflag == 1, Print["4ab"]];
Clear[temp1];
temp1 = ConstantArray[{}, 7];
Do [temp1 = {i, Log[extenlst]};, {i, 1, 7}];
nlm1 =
NonlinearModelFit[temp1,
nlm1pars[a2, f7, c2, d2, x], {a2, f7, c2, d2}, x,
MaxIterations -> 30000];
If[debugflag == 1, Print["4ac"]];
If [debugflag == 1,
Print["4e"]];
;

Clear[temp1, temp2, count];
temp2 = {};
count = 0;
temp1 = 0;
```

```
While[Re[Exp[nlm1[count + 8]]] <= extendflag,
If[Re[Exp[nlm1[count + 8]]] > xb,
AppendTo[temp2, Re[Exp[nlm1[count + 8]]]]; temp1 = temp1 + 1;];
count = count + 1;
];

Print["Extended Magnitudes are: ", temp2];
If[debugflag == 1, Print["4f"]];

(*extenlst = Join[1/temp2,extenlst];*)
extenlst = Join[extenlst, temp2];
pdlist = N[Join[pdlist, ConstantArray[1.0, count]]];
];
If [extendflag == 0, temp1 = 7];
order = Ordering[extenlst];
extenlst = N[Sort[extenlst]];
pdlist = N[pdlist];
Print[temp1];
Print[extenlst];
Print[pdlist];
Print[extenlst];
Print[pdlist];
N1 = Length[extenlst];
xb = extenlst;
xa = extenlst;

If[debugflag == 1, Print["4g"]];

f = ConstantArray[{}, N5];
func1 = ConstantArray[{}, N1];
outlst = ConstantArray[0, N5];
pdflist = ConstantArray[0, N5];
densitylist = ConstantArray[{}, N1];
density = ConstantArray[0, N5];

Clear[f2];
Do [func1 = {i, (SetPrecision[extenlst, 30])}, {i, 1,
```

```
N1}];
y = Interpolation[func1];
h0 = N[(N1 - 1)/(N5 - 1)];
Do [
f = {N[h0*i + 1], N[y[h0*i + 1]]}, {i, 0, N5 - 1}];
f2 = Interpolation[f];
Do[densitylist = {i, pdlist};, {i, 1, N1}];
y2 = Interpolation[densitylist];
Do[density = N[y2[h0*i + 1]];, {i, 0, N5 - 1}];
temp2 = Derivative[1][f2];
Do [If[temp2[h0*i + 1] == 0,
If[i == 0,
pdflist =
density/N[temp2[h0 + 1]] + 10^(-200);,
pdflist = pdflist + 10^(-200);];,
pdflist = density/N[temp2[h0*i + 1]]], {i, 0,
N5 - 1}];

Do [outlst = N[y[(i*h0 + 1)]], {i, 0, N5 - 1}];
Do[If[pdflist < 0, pdflist = 0], {i, 1, N5}];
{sumval, intlist} = getIntegrate2[pdflist, 1, 1, N5];
pdflist = pdflist/sumval;
intlist = intlist/sumval;
If[debugflag == 1, Print["End GetPDF5"]];
{xa, xb, extenlst, intlist, pdflist, outlst}]
ClearAll[getIntegrate2];

getIntegrate2[datapts_?(VectorQ[#, NumberQ] &), k_?NumberQ,
a_?NumberQ, b_?NumberQ] :=
Module[{N1 = Length[datapts], coeffs = {14, 32, 12, 32}, h1,
intlist, sumval, j, temp1}, h1 = ((b - a)/(N1 - 1)*2/45);
intlist = ConstantArray[0, N1];
intlist = N[7*datapts^k*h1]; sumval = intlist;
j = 1;
While[j <= N1 - 2,
temp1 = coeffs*datapts^k*h1;
```

```
sumval = sumval + temp1; intlist = N[sumval]; j += 1];
sumval = sumval + 7*datapts^k*h1;
intlist = N[sumval];
{N[sumval], intlist}];
ClearAll[getMultiply2];

getMultiply2[datapts_List, k_?NumberQ, a_?NumberQ, b_?NumberQ] :=
Module[{N1 = Length[datapts], intlist, mulval, j, temp1},
intlist = {datapts^k}; mulval = intlist;
j = 1; While[j <= N1 - 2, temp1 = datapts^k;
mulval = N[mulval + temp1 - mulval*temp1];
intlist = Append[intlist, mulval]; j += 1];
mulval = N[mulval + datapts^k - mulval*datapts^k];
intlist = Append[intlist, mulval];
N[intlist]]

ClearAll[getDist];(*new definition from 08/10/2017*)
getDist[lst_List] :=
Module[{probdist, phasedistpdf, phasedistcdf, count, magvals,
sumval1, indx, order, L, lst2, temp, flag, i, j, k, lst1, newlist,
p, newmaglist, sdifflist, sumval, lst3, temp1, f2, func1, y,
numpts, N2, xa, xb, h0, temp2, phaselistpdf, indexf, ratio,
phaselistcdf, outmags, lst4, pdlist, N1, h1, k1, diffs, N5},

Print["Start getDist"];
lst1 = lst;
temp = matlabFind1[lst1, Greater, 0];
lst2 = lst1;
lst3 = Sort[lst2];
L = Length[lst2];
flag = 0;
Do[If[lst2 != NumberQ || lst2 <= 0 ||
lst2 == Infinity, flag = 1;], {j, 1, L}];
lst4 = {}; pdlist = {};
j = 1; While [j < L, k = 1;
While[j < L && lst3 == lst3, k = k + 1; j = j + 1;];
```

```
AppendTo[pdlist, k]; AppendTo[lst4, lst3]; j = j + 1];
If[lst3 != lst3, AppendTo[pdlist, 1];
AppendTo[lst4, lst3];];
N1 = Length[lst4];

If[flag == 1,
Print["GetDist:Illegal, some of the data values are not numbers \
are not >0 or are infinity: ", lst3]; Exit[];];
order = Ordering[lst2];
sdifflist = Sort[Differences[order]];

L = L - 1;
newlist = {};
j = 1;
i = -L + 1; diffs = {};
While[i <= L - 1 && j <= L, k = 0;
While[j <= L && sdifflist == i, k = k + 1;
j = j + 1];
If[k > 0, AppendTo[newlist, k]; AppendTo[diffs, i]];
i = i + 1];

numpts = 16384;
N2 = 2^(Floor[Log[Length[phasedistpdf]]/Log[2.0] + 1]);
If[N2 < numpts, N2 = numpts];
N5 = Length[newlist];

Clear[temp];
temp = ConstantArray[{}, N5];
Do[temp = {i,
SetPrecision[N[newlist], 30]/Total[newlist]};, {i, 1, N5}];

y = Interpolation[temp];
h0 = N[(N5 - 1)/(N2 - 1)];
phaselistpdf = ConstantArray[SetPrecision[0.0, 30], N2];
Do[phaselistpdf = N[Evaluate[y[h0*i + 1]]];, {i, 0,
N2 - 1}];
```

```
Clear[temp];
temp = ConstantArray[{}, N5];
Do[temp = {i, SetPrecision[diffs, 30]};, {i, 1, N5}];
Clear[y];
y = Interpolation[temp];
indexf = ConstantArray[0, N2];
Do[indexf = N[Evaluate[y[h0*i + 1]]];, {i, 0, N2 - 1}];
f2 = ConstantArray[{}, N1];
Do[f2 = {i, SetPrecision[lst4, 30]}, {i, 1, N1}];
Clear[y];
y = Interpolation[f2];

outmags = ConstantArray[0, N2];
h1 = N[(N1 - 1)/(N2 - 1)];
Do [
outmags = N[Evaluate[y[h1*i + 1]]], {i, 0, N2 - 1}];

Do[If[phaselistpdf <= 0, phaselistpdf = 10^(-200);], {i,
1, N2}];
Clear[sumval, phaselistcdf];
{sumval, phaselistcdf} = getIntegrate2[phaselistpdf, 1, 1, N2];

phaselistpdf = N[phaselistpdf/sumval];
phaselistcdf = N[phaselistcdf/sumval];
ratio = N[N2/L];
Print["End getDist"];
{phaselistpdf, phaselistcdf, newlist, indexf, outmags, ratio}];

+
ClearAll[generateMagPhase];
generateMagPhase[ratios_List, pdflist_List, outlst_List, np_Integer,
steps_Integer,
debugflag_Integer] /; (debugflag == 1 || debugflag == 0) :=
Module[{ft, sigma, psi, mu, x, nlm, h, h1, dB, pdB, sumval, newlist,
magvals, temp1, count, N3, probdist, i, j, simvals, k, x0,
indxvals, indx, g1list, g2list, rv, k1, p2, futureflag, mag, p1,
```

```
phasedistcdf, diff, temp, phaselistpdf, ratio, indexf, outmags, k2,
p11, phaselistcdf, x2, weights, p3, magprob, magprobdist, maglist,
p, sigma1, psi1, rv1, mu1, k3, N5},
Print["Start GenerateMagPhase:", DateString[]];
{phaselistpdf, phaselistcdf, newlist, indexf, outmags, ratio} =
getDist[ratios];

If[debugflag == 1, Print["****1"]];

If[debugflag == 1, Print["****2"]];

N5 = Length[outmags];

g2list = ConstantArray[0, N5];
i = 1; k = 1;
While[i <= N5, x0 = N[(i - 1)/(N5 - 1)];
While[(k <= N5 && phaselistcdf < x0), k = k + 1];
If[k > 1, g2list = k - 1, g2list = 1.];
i = i + 1];
dB = ConstantArray[SetPrecision[0.0, 30], {np, steps}];
pdB = ConstantArray[0.0, {np, steps}];

Do[k1 = Floor[RandomReal[]*(N5 - 1) + 1.5];

Do[

k = g2list;
diff = indexf;
k2 = Round[k1 + ratio*diff];
While[k2 < 1 || k2 > N5, k1 = Floor[RandomReal[]*(N5 - 1) + 1.5];
k = g2list; diff = indexf;
k2 = Round[k1 + ratio*diff];];

k1 = k2;

mag = outmags;
p2 = phaselistpdf;
```

```
If[p2 <= 0, p2 = 10^(-200)];

dB = mag;
pdB = p2;, {j, 1, steps}];, {i, 1, np}];
Print["End GenerateMagPhase:", DateString[]];
{outlst, pdflist, outmags, phaselistpdf, ratio, indexf, dB, pdB}]

ClearAll[generateTemperatures]; (* 11/6/2012 and 12/10/2012 and \
12/11/2012 where s0 is initial price *)

generateTemperatures[dB_List?MatrixQ, s0_Real] :=
Module[{temperatures, i, j, np, steps},
{np, steps} = Dimensions[dB];
temperatures = ConstantArray[0, {np, steps}];
Print["Start getTemperatures"];
temperatures = s0*dB;
Do[
Do[temperatures = temperatures*dB, {j,
2, steps}], {i, 1, np}];
(*Do[Do[temperatures=temperatures+coeffs*j,{j,1,
steps}],{i,1,np}];*)

temperatures]

ClearAll[getPD];

getProbDist[
filename_String: \
"C:\\Users\\Chondrally\\Documents\\Matlab\\query.csv", label_String,
loopfactor_Integer, percentthreshold_, IndicatorPaths_Integer,
extendflagmags_: 1, extendflagtimes_: 0, includemagquakes_,
excludetimequakes_, earthquakecutoff_List,
earthquakespecialcutoff_Real, daysAheadCutoff_Integer,
StartEarthQuakeNumber_Integer, FutureDate_, debugflag_Integer: 0] :=
Module[{i, j, k, numbersteps, temp1, temp2, coeffs, xa3, xb3,
ratios3, extenlst3, intlist3, pdflist3, outlst3, dB3, pdB3,
phaselistpdf3, xa4, xb4, ratios4, extenlst4, intlist4, pdflist4,
```

```
outlst4, dB4, pdB4, phaselistpdf4, str, count, mags, times,
probdist, timej, temperatures3, temperatures4, deTrendflag,
timestamp1, timestamp2, plotFutureARMA, str2, labels, plot1,
plotFouriermags, plotFourierTimes, timesFourier, magsFourier, data,
filepath, temp, maxval, plotFourier, mintemp1, totalsteps, a1, b1,
c1, mags2, times2, plotTestData, flag, timek, temp3, plotCombined,
plotCombined2, plotQMtimes, plotQMmags, plot2, a, b, c, x, nlm,
probdist2, nlm2, plotFutureARMA2, a2, b2, c2, timeMagnitude, time1,
time2, sumval1, sumval2, daysmean, daysmean3, plot7, plot8,
magprobdist1, magprobdist2, p1, p2, daysmean1, count5, daysmean2,
probdista, probdistcdf, probdistb, ymin, Magnitudes,
timeMagnitudes, count1, count2, count3, count4, count1a, count2a,
count3a, count4a, probval, sdcount1, sdcount2, sdcount3, sdcount4,
count6, x0, k1, N5, g2list, sd, N1, count7, N3, k3, quakecounts,
quaketimes, probval2, InitialQuakeCount, ActualDaysfromprob, ratio,
indexf, outlst, pdflist},

Print["label = ", label];
Print["filename = ", filename];
Print["FutureDate = ", FutureDate, " years forecast"];
Print["IndicatorPaths = ", IndicatorPaths];
Print["extendflagmags = ", extendflagmags];
Print["extendflagtimes = ", extendflagtimes];
Print["earthquakespecialcutoff = ", earthquakespecialcutoff];
Print["earthquakecutoff = ", earthquakecutoff];
Print["daysAheadCutoff = ", daysAheadCutoff];
Print["StartEarthQuakeNumber = ", StartEarthQuakeNumber];
Print["includemagquakes = ", includemagquakes];
Print["excludetimequakes = ", excludetimequakes];
Print["debugflag= ", debugflag];

timestamp1 = DateString[];
Print["timestamp1 = ", timestamp1];
deTrendflag = 1;

time2 = N[FutureDate*365.2422];
```

```
filepath = FileNameJoin[{NotebookDirectory[], filename}];
data = Rest[Import[filepath]];
count = Length[data];
If [count < 1, Print["No DATA"]; Exit[];];
Print["Number of Earthquakes in file = ", count];
timeMagnitude = data /. {x_, y_} :> {N[AbsoluteTime[x]/60/60/24], y};

If[StartEarthQuakeNumber < count,

count = count - StartEarthQuakeNumber + 1;
totalsteps = count;

times = ConstantArray[0, count - 1];
mags = ConstantArray[0, count];
maxval = -1;
temp = 0;
times =
Table[N[(timeMagnitude - timeMagnitude)], {i,
StartEarthQuakeNumber, count - 1}];
mags =
Table[timeMagnitude, {i, StartEarthQuakeNumber + 1,
count}];
Clear[temp1, temp1, temp3];
temp1 = matlabFind1[times, GreaterEqual, excludetimequakes];
temp2 = matlabFind1[mags, GreaterEqual, includemagquakes];
temp3 = Union[temp1, temp2];
times = times;
mags = mags;
count = Length[temp3];
Print["number of valid earthquakes in file, time difference > ",
ToString[excludetimequakes], " days and greater than magnitude ",
ToString[includemagquakes], "= ", count];
time1 = timeMagnitude;
numbersteps = Round[time2];
Clear[temp1, temp2];
temp1 = N[Mean[times]];
```

```
Print["Mean number of days per earthquake = ", temp1];
j = 1; temp2 = 0;
daysmean = Floor[N[numbersteps/temp1] + 500];
Print["number of days in 30 years is : ", time2];
Print["Average number of days per path : ",
N[(daysmean - 500)*temp1]];
Print ["number of significant earthquakes till ", FutureDate,
"years from last earthquake in file is : ", daysmean - 500];,

Print["not enough data in file for StartEarthQuakeNumber"];
Exit[];
];
If [debugflag == 1, Print[times]; Print[mags];];

If[debugflag == 1, Print[Length[mags], Length[times]]];
If [debugflag == 1, Print["**1"];];
If[debugflag == 1, Print[mags]; Print[times];];

Clear[temp1, temp2];
{xa4, xb4, ratios4, extenlst4, intlist4, pdflist4, outlst4} =
getData[times, 16384, extendflagtimes, debugflag];
N1 = 1000; timej = ConstantArray[0, N1]; N5 = Length[intlist4];
g2list = ConstantArray[0, N5];
i = 1; k = 1;
While[i <= N5, x0 = N[(i - 1)/(N5 - 1)];
While[(k <= N5 && intlist4 < x0), k = k + 1];
If[k > 1, g2list = (k - 1), g2list = 1.];
i = i + 1];
Do[timek = 0; j = 1;
While[timek <= Round[time2],
k1 = Floor[RandomReal[]*(N5 - 1) + 1.5]; k = g2list;
timek = timek + outlst4; j = j + 1;];
timej = j - 1;, {i, 1, N1}];

sd = N[Sqrt[Total[(timej - Mean[timej])^2]/(N1 - 1)]];
numbersteps = Round[Mean[timej] + percentthreshold*sd];
```

```
Print["sample size numbersteps =", N1];
Print["mean numbersteps =", N[Mean[timej]]];
Print["standard deviation numbersteps =", sd];
Print["Estimated numbersteps =", numbersteps];
Clear[timej, timek, g2list];
{xa3, xb3, ratios3, extenlst3, intlist3, pdflist3, outlst3} =
getData[mags, 16384, extendflagmags, debugflag];
probdist = ConstantArray[SetPrecision[0.0, 30], Round[time2] + 1];
Clear[count1, timej, timek];
N3 = loopfactor;
count7 = ConstantArray[SetPrecision[0.0, 30], {N3, IndicatorPaths}];
count1 =
count2 =
ConstantArray[
SetPrecision[0.0, 30], {Length[earthquakecutoff], 4,
IndicatorPaths}];
quakecounts = ConstantArray[0, {4, Length[earthquakecutoff]}];
quaketimes = ConstantArray[0, {4, Length[earthquakecutoff]}];
InitialQuakeCount = ConstantArray[0, Length[earthquakecutoff]];
probval = probval2 = ConstantArray[0, {4, Length[earthquakecutoff]}];
ActualDaysfromprob = ConstantArray[0, Length[earthquakecutoff]];
Do[
Print["Country = ", label];
Print["k3=", k3, ", IndicatorPaths= ", IndicatorPaths,
", Totalk3 = ", N3, " total paths= ", N[N3*IndicatorPaths]];

count5 = count6 = ConstantArray[0, IndicatorPaths];
(*{coeffs,temp1,temp2}=getPercentErrorLinear[mags,debugflag];
mintemp1 = Min[temp1];
temp1 = temp1 - mintemp1 +1;*)

Clear[plot1, magsFourier];
(*magsFourier = ConstantArray[0,{3,numbersteps}];
{plot1,magsFourier}=MusicFinal[mags,numbersteps,mags,Length[
mags],0,debugflag];*)
If[debugflag == 1, Print["*****2"]];
```

```
If[debugflag == 1,
Print[Dimensions[ratios3], Dimensions[intlist3],
Dimensions[pdflist3], Dimensions[outlst3]]];
Clear[magprobdist1, p1, Magnitudes, pdB3, phaselistpdf3, outlst,
pdflist, ratio, indexf];
{outlst, pdflist, magprobdist1, phaselistpdf3, ratio, indexf,
Magnitudes, pdB3}
= generateMagPhase[mags, pdflist3, outlst3, IndicatorPaths,
numbersteps, debugflag];

Clear[plot2, timesFourier];
(*timesFourier = ConstantArray[0,{3,numbersteps}];

{plot2,timesFourier}=MusicFinal[times,numbersteps,times,
Length[times],0,debugflag];*)
If[debugflag == 1, Print["*****2"]];
If[debugflag == 1,
Print[Dimensions[ratios4], Dimensions[intlist4],
Dimensions[pdflist4], Dimensions[outlst4]]];
Clear[magprobdist2, p2, timeMagnitudes, pdB4, phaselistpdf4, ratio,
indexf, outlst, pdflist];
{outlst, pdflist, magprobdist2, phaselistpdf4, ratio, indexf,
timeMagnitudes, pdB4} =
generateMagPhase[times, pdflist4, outlst4, IndicatorPaths,
numbersteps, debugflag];

Clear[temp];

If [debugflag == 1, Print["**1"]];
Print["ActualDaysfromprob before: ", ActualDaysfromprob];
Do[
If[k3 == 1,
quaketimes = {N[time2/FutureDate], N[time2/3],
N[time2*2/3], N[time2]};];

If[k3 == 2,
InitialQuakeCount =
```

```
N[Mean[count1]];
count1 =
ConstantArray[0, IndicatorPaths];
count2 =
ConstantArray[0, IndicatorPaths];

If[Positive[InitialQuakeCount],
ActualDaysfromprob = N[365.2422/InitialQuakeCount]];
If [ActualDaysfromprob > time2 || !
Positive[InitialQuakeCount],
ActualDaysfromprob = time2;];
quaketimes = {N[ActualDaysfromprob],
N[time2/3], N[time2*2/3], N[time2]};
quakecounts = 0;];
If[k3 > 2, quakecounts = 0;
Clear[temp1];
temp1 = N[Mean[count1]];
If[Positive[temp1],
ActualDaysfromprob =
N[(2.0*ActualDaysfromprob/temp1 +
ActualDaysfromprob)/3.0];]
If[! Positive[temp1] || ActualDaysfromprob > time2,
ActualDaysfromprob = time2;];
count1 =
ConstantArray[0, IndicatorPaths];
count2 =
ConstantArray[0, IndicatorPaths];
quaketimes = {N[ActualDaysfromprob],
N[time2/3], N[time2*2/3], N[time2]};];, {k1, 1,
Length[earthquakecutoff]}];
Print["ActualDaysfromprob After = ", ActualDaysfromprob];
Print["InitialQuakeCount = ", InitialQuakeCount];
Print["numbersteps =", numbersteps];
Do[Clear[temp1]; timej = 0; timek = 0; j = 1;
While[timej <= time2 && j <= numbersteps,
temp1 = timeMagnitudes; timej = N[timej + temp1];
```

```
timek = Round[timej];
Do[
Do [
If[timej <= quaketimes,
quakecounts = quakecounts + 1;];, {k1, 1,
4}];, {k, 1, Length[earthquakecutoff]}];

Do[

If[Magnitudes >= earthquakecutoff,
Do[

If[timej <= quaketimes,
count1 = count1 + 1;
count2 =
count2 + pdB3;];, {k1, 1, 4}]];, {k,
1, Length[earthquakecutoff]}];
If [Magnitudes >= earthquakespecialcutoff && timek >= 1 &&
timek <= Round[time2],
probdist = probdist + pdB3];
j = j + 1;]; count5 = j - 1;
count6 = timej - temp1;, {i, 1, IndicatorPaths}];
Print["Mean of count5 = ", N[Mean[count5]],
": Total count5 = ", N[Total[count5]]];
Print["Mean of count6 = ", N[Mean[count6]]];

count7 = count7 + count5;

Print["probdist2 calulated"];

Clear[count1a, sdcount1];
Print["Country = ", label];
count1a =
sdcount1 = ConstantArray[0, {Length[earthquakecutoff], 4}];
Do[Do[If[k1 == 1,
count1a =
N[Mean[count1]];,
```

```
count1a =
N[Mean[count1]/k3];];

sdcount1 =
N[Sqrt[Total[(count1/k3 -
count1a)^2]/(IndicatorPaths - 1)]];
If [k3 > 1 && k1 > 1,
probval =
N[Mean[count2]/
k3/(quakecounts/k3/IndicatorPaths)];
, probval =
N[Mean[count2]/(quakecounts/
IndicatorPaths)];];
If[k3 > 1 && k1 > 1,
probval2 =
N[probval*(quakecounts/(k3 - 1)/IndicatorPaths)/
quaketimes];,
probval2 =
N[probval*(quakecounts/IndicatorPaths)/
quaketimes];;];
Print["magnitude cutoff = ", earthquakecutoff,
", years = ", N[quaketimes/365.2422],
", Probability= ", probval, " per quake or ",
probval2, " per day, count = ", N[count1a],
", sd = ", sdcount1,
", mean number of quakes in time period is: ",
If[k1 == 1,
If[k3 == 1, N[quakecounts/IndicatorPaths],
N[quakecounts/(k3 - 1)/IndicatorPaths]],
N[quakecounts/k3/IndicatorPaths]],
", Number of days in period = ", quaketimes];, {k1,
1, 4}];, {k, 1, Length[earthquakecutoff]}];

, {k3, 1, N3}];
probdist = N[probdist/Total[count7, 2]];
```

```
Clear[temp];
temp = ConstantArray[{}, Length[phaselistpdf3]];
Do [temp = {magprobdist1, phaselistpdf3};, {i, 1,
Length[phaselistpdf3]}];
Clear[temp2, plot7];
temp2 =
StringJoin["Probability vs. Earthquake Magnitude for ", label];
plot7 =
ListLinePlot[temp, PlotRange -> Automatic, PlotStyle -> Blue,
Joined -> True, AxesLabel -> {"Magnitude", "Probability"},
PlotLabel -> temp2, GridLines -> Automatic, RotateLabel -> True];
If[debugflag == 1, Print["*****3"]];

Clear[temp];
temp = ConstantArray[{}, Length[phaselistpdf4]];
Do [temp = {magprobdist2, phaselistpdf4};, {i, 1,
Length[phaselistpdf4]}];
Clear[temp2, plot8];
temp2 =
StringJoin["Probability vs. Earthquake time Differences for ",
label];
plot8 =
ListLinePlot[temp, PlotRange -> Automatic, PlotStyle -> Blue,
Joined -> True, AxesLabel -> {"TimeMagnitudes", "Probability"},
PlotLabel -> temp2, GridLines -> Automatic, RotateLabel -> True];
If[debugflag == 1, Print["*****3"]];

Print["Standard Deviations calculated"];

If[debugflag == 1, Print["**1"]];

If[debugflag == 1, Print["**2"]];

Clear[temp];
temp = ConstantArray[{}, Round[time2] + 1];
Do[temp = {i, N[probdist]};, {i, 1, Round[time2] + 1}];
Print["plot array calculated"];
```

```
Clear[temp2];
temp2 =
StringJoin["Probability of greater than ",
ToString[earthquakespecialcutoff], " Richter Earthquake in ",
label, " between ", data, " and ", ToString[FutureDate],
"years from then"];
plotFutureARMA =
ListLinePlot[temp, PlotRange -> Automatic, PlotStyle -> Blue,
Joined -> True, AxesLabel -> {"days ahead", "Probability"},
PlotLabel -> temp2, GridLines -> Automatic, RotateLabel -> True];
Clear[temp2];

Print["plot made"];

Print["Number of Earthquakes in file = ", count];
Print ["number of earthquakes till ", FutureDate,
"years from last earthquake in file is : ", daysmean];

timestamp2 = DateString[];
Print["timestamp2 = ", timestamp2];
{magprobdist1, magprobdist2, phaselistpdf3, phaselistpdf4, probdist,
count1, sdcount1,
Framed[GraphicsColumn[{plot7, plot8, plotFutureARMA},
ImageSize -> {700, 1500}]]}]

filename = "queryNewZealand.csv"; (* Select a .csv filename on your \
drive in the same directory as this program resides*)

label = "New Zealand";
debugflag = 0; (* if you want to see the progress of the evaluation, \
set debugflag =1 *)

FutureDate = 30; (*the future date from the last earthquake in the \
file to this date in the future will calculate the probability of an \
earthquake greater than earthquakemagnitudecutoff between those two \
dates*)
StartEarthQuakeNumber = 1;(* if you want to omit some of the data in \
```

```
the file, set this number higher than 1 to include the data from \
that earthquakenumber to the end of the file *)
daysAheadCutoff = 10;(* the number of days into the future for the \
second probability plot, to see what the probability is of an \
earthquake greater than than the earthquakemagnitudecutoff within \
that many days*)
paths = 2000;(* 1000 is 0.1% error and 4000 is 0.025% error and 10000 \
is .01% error. number of paths into future to calculate for \
indicatorpaths and for main prediction *)
earthquakemagnitudecutoff = {5.0, 5.5, 6.0, 6.5, 7.0, 7.5, 8.0, 8.5,
9.0};
earthquakespecialcutoff = 7.0; (* this is the cutoff threshold so \
that only earthquakes above this threshold are used to report the \
plot *)
extendflagmags = 10.0;(*extend the earthquake magnitudes ratio \
distribution *)
extendflagtimes = .1;(* extend the earthquake time differences ratio \
distribution *)

excludetimequakes = 0; (* mean number of days threshold for \
earthquakes, exclude earthquakes with a time difference less than \
this *)
includemagquakes = 2.5; (* include all earthquakes with magnitude of \
this and higher *)

percentthreshold = .3;
loopfactor = 10;

getProbDist[filename, label, loopfactor, percentthreshold, paths, \
extendflagmags, extendflagtimes, includemagquakes, excludetimequakes, \
earthquakemagnitudecutoff, earthquakespecialcutoff, daysAheadCutoff, \
StartEarthQuakeNumber, FutureDate, debugflag]
```

Printed in the United States
By Bookmasters